EXCLUSIVE
EMPLOYMENT ADVICE

CARPENTRY & JOINERY

LEVEL 1 DIPLOMA

PATRICK JOSEPH CLANCY
STEVEN LEAVERLAND

Nelson Thornes

Published in 2013 by:
Nelson Thornes Ltd
Delta Place
27 Bath Road
CHELTENHAM
GL53 7TH
United Kingdom

13 14 15 16 17 / 10 9 8 7 6 5 4 3 2 1

A catalogue record for this book is available from the British Library

ISBN 978 1 4085 2125 0

Cover photograph: tuja66/Fotolia

Page make-up by GreenGate Publishing Services, Tonbridge, Kent

Printed in Croatia by Zrinski

Note to learners and tutors

This book clearly states that a risk assessment should be undertaken and the correct PPE worn for the particular activities before any practical activity is carried out. Risk assessments were carried out before photographs for this book were taken and the models are wearing the PPE deemed appropriate for the activity and situation. This was correct at the time of going to print. Colleges may prefer that their learners wear additional items of PPE not featured in the photographs in this book and should instruct learners to do so in the standard risk assessments they hold for activities undertaken by their learners. Learners should follow the standard risk assessments provided by their college for each activity they undertake which will determine the PPE they wear.

CONTENTS

INTRODUCTION

About this book

This book has been written for the Cskills Awards Level 2 Diploma in Bricklaying. It covers all the units of the qualification, so you can feel confident that your book fully covers the requirements of your course.

This book contains a number of features to help you acquire the knowledge you need. It also demonstrates the practical skills you will need to master to successfully complete your qualification. We've included additional features to show how the skills and knowledge can be applied to the workplace, as well as tips and advice on how you can improve your chances of gaining employment.

The features include:

* chapter openers which list the learning outcomes you must achieve in each unit

* key terms that provide explanations of important terminology that you will need to know and understand

* Did you know? margin notes to provide key facts that are helpful to your learning

* practical tips to explain facts or skills to remember when undertaking practical tasks

* Reed tips to offer advice about work, building your CV and how to apply the skills and knowledge you have learnt in the workplace

* case studies that are based on real tradespeople who have undertaken apprenticeships and explain why the skills and knowledge you learn with your training provider are useful in the workplace

* practical tasks that provide step-by-step directions and illustrations for a range of projects you may do during your course

* Test yourself multiple choice questions that appear at the end of each unit to give you the chance to revise what you have learnt and to practise your assessment (your tutor will give you the answers to these questions).

Further support for this book can be found at our website, www.planetvocational.com/subjects/build

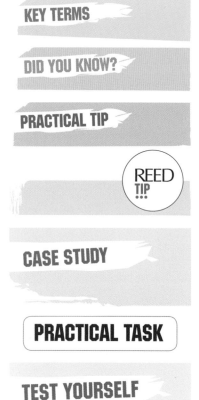

KEY TERMS

DID YOU KNOW?

PRACTICAL TIP

REED TIP
•••

CASE STUDY

PRACTICAL TASK

TEST YOURSELF

Planet Vocational

CONTRIBUTORS TO THIS BOOK

Reed Property & Construction

Reed Property & Construction specialises in placing staff at all levels, in both temporary and permanent positions, across the complete lifecycle of the construction process. Our consultants work with most major construction companies in the UK and our clients are involved with the design, build and maintenance of infrastructure projects throughout the UK.

Expert help

As a leading recruitment consultancy for mid–senior level construction staff in the UK, Reed Property & Construction is ideally placed to advise new workers entering the sector, from building a CV to providing expertise and sharing our extensive sector knowledge with you. That's why, throughout this book, you will find helpful hints from our highly experienced consultants, all designed to help you find that first step on the construction career ladder. These tips range from advice on CV writing to interview tips and techniques, and are all linked in with the learning material in this book.

Work-related advice

Reed Property & Construction has gained insights from some of our biggest clients – leading recruiters within the industry – to help you understand the mind-set of potential employers. This includes the traits and skills that they would like to see in their new employees, why you need the skills taught in this book and how they are used on a day to day basis within their organisations.

Getting your first job

This invaluable information is not available anywhere else and is all geared towards helping you gain a position with an employer once you've completed your studies. Entry level positions are not usually offered by recruitment companies, but the advice we've provided will help you to apply for jobs in construction and hopefully gain your first position as a skilled worker.

CONTRIBUTORS TO THIS BOOK

The case studies in this book feature staff from Laing O'Rourke and South Tyneside Homes.

Laing O'Rourke is an international engineering company that constructs large-scale building projects all over the world. Originally formed from two companies, John Laing (founded in 1848) and R O'Rourke and Son (founded in 1978) joined forces in 2001.

At Laing O'Rourke, there is a strong and unique apprenticeship programme. It runs a four-year 'Apprenticeship Plus' scheme in the UK, combining formal college education with on-the-job training. Apprentices receive support and advice from mentors and experienced tradespeople, and are given the option of three different career pathways upon completion: remaining on site, continuing into a further education programme, or progressing into supervision and management.

The company prides itself on its people development, supporting educational initiatives and investing in its employees. Laing O'Rourke believes in collaboration and teamwork as a path to achieving greater success, and strives to maintain exceptionally high standards in workplace health and safety.

South Tyneside Council's
Housing Company

South Tyneside Homes was launched in 2006, and was previously part of South Tyneside Council. It now works in partnership with the council to repair and maintain 18,000 properties within the borough, including delivering parts of the Decent Homes Programme.

South Tyneside Homes believes in putting back into the community, with 90 per cent of its employees living in the borough itself. Equality and diversity, as well as health and wellbeing of staff, is a top priority, and it has achieved the Gold Status Investors in People Award.

South Tyneside Homes is committed to the development of its employees, providing opportunities for further education and training and great career paths within the company – 80 per cent of its management team started as apprentices with the company. As well as looking after its staff and their community, the company looks after the environment too, running a renewable energy scheme for council tenants in order to reduce carbon emissions and save tenants money.

The apprenticeship programme at South Tyneside Homes has been recognised nationally, having trained over 80 young people in five main trade areas over the past six years. One of the UK's Top 100 Apprenticeship Employers, it is an Ambassador on the panel of the National Apprentice Service. It has won the Large Employer of the Year Award at the National Apprenticeship Awards and several of its apprentices have been nominated for awards, including winning the Female Apprentice of the Year for the local authority.

Chapter 1

Unit CSA–L1Core01

HEALTH, SAFETY AND WELFARE IN CONSTRUCTION AND ASSOCIATED INDUSTRIES

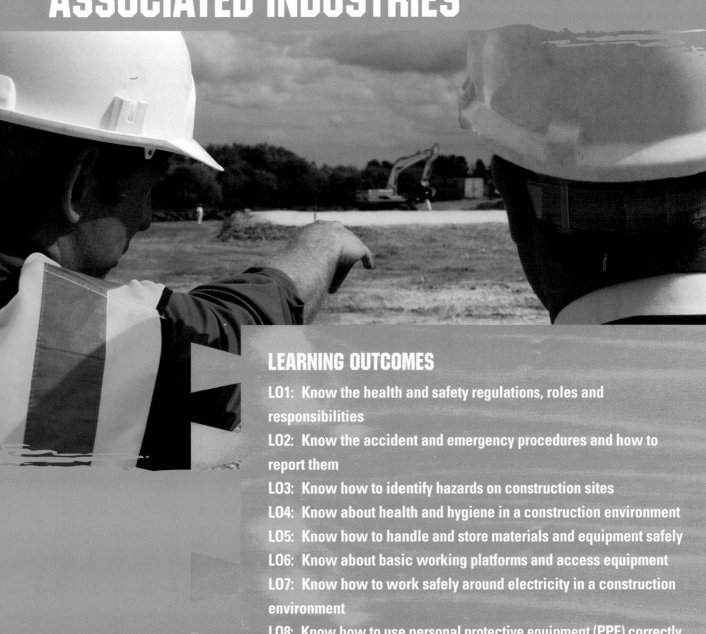

LEARNING OUTCOMES

LO1: Know the health and safety regulations, roles and responsibilities

LO2: Know the accident and emergency procedures and how to report them

LO3: Know how to identify hazards on construction sites

LO4: Know about health and hygiene in a construction environment

LO5: Know how to handle and store materials and equipment safely

LO6: Know about basic working platforms and access equipment

LO7: Know how to work safely around electricity in a construction environment

LO8: Know how to use personal protective equipment (PPE) correctly

LO9: Know the fire and emergency procedures

LO10: Know about signs and safety notices

INTRODUCTION

The aim of this chapter is to:

* help you to source relevant safety information
* help you to use the relevant safety procedures at work.

HEALTH AND SAFETY REGULATIONS, ROLES AND RESPONSIBILITIES

The construction industry can be dangerous, so keeping safe and healthy at work is very important. If you are not careful, you could injure yourself in an accident or perhaps use equipment or materials that could damage your health. Keeping safe and healthy will help ensure that you have a long and injury-free career.

Although the construction industry is much safer today than in the past, more than 2,000 people are injured and around 50 are killed on site every year. Many others suffer from long-term ill-health such as deafness, spinal damage, skin conditions or breathing problems.

Key health and safety legislation

Laws have been created in the UK to try to ensure safety at work. Ignoring the rules can mean injury or damage to health. It can also mean losing your job or being taken to court.

The two main laws are the Health and Safety at Work etc. Act (HASAWA) and the Control of Substances Hazardous to Health Regulations (COSHH).

The Health and Safety at Work etc. Act (HASAWA) (1974)
This law applies to all working environments and to all types of worker, sub-contractor, employer and all visitors to the workplace. It places a duty on everyone to follow rules in order to ensure health, safety and welfare. Businesses must manage health and safety risks, for example by providing appropriate training and facilities. The Act also covers first aid, accidents and ill health.

Reporting of Injuries, Diseases and Dangerous Occurrences Regulations (RIDDOR) (1995)
Under RIDDOR, employers are required to report any injuries, diseases or dangerous occurrences to the Health and Safety Executive (HSE). The regulations also state the need to maintain an accident book.

KEY TERMS

HASAWA

– the Health and Safety at Work etc. Act outlines your and your employer's health and safety responsibilities.

COSHH

– the Control of Substances Hazardous to Health Regulations are concerned with controlling exposure to hazardous materials.

DID YOU KNOW?

In 2011 to 2012, there were 49 fatal accidents in the construction industry in the UK. (*Source* HSE, www.hse.gov.uk)

KEY TERMS

HSE

– the Health and Safety Executive, which ensures that health and safety laws are followed.

Accident book

– this is required by law under the Social Security (Claims and Payments) Regulations 1979. Even minor accidents need to be recorded by the employer. For the purposes of RIDDOR, hard copy accident books or online records of incidents are equally acceptable.

Control of Substances Hazardous to Health (COSHH) (2002)

In construction, it is common to be exposed to substances that could cause ill health. For example, you may use oil-based paints or preservatives, or work in conditions where there is dust or bacteria.

Employers need to protect their employees from the risks associated with using hazardous substances. This means assessing the risks and deciding on the necessary precautions to take.

Any control measures (things that are being done to reduce the risk of people being hurt or becoming ill) have to be introduced into the workplace and maintained; this includes monitoring an employee's exposure to harmful substances. The employer will need to carry out health checks and ensure that employees are made aware of the dangers and are supervised.

Control of Asbestos at Work Regulations (2012)

Asbestos was a popular building material in the past because it was a good insulator, had good fire protection properties and also protected metals against corrosion. Any building that was constructed before 2000 is likely to have some asbestos. It can be found in pipe insulation, boilers and ceiling tiles. There is also asbestos cement roof sheeting and there is a small amount of asbestos in decorative coatings such as Artex.

Asbestos has been linked with lung cancer, other damage to the lungs and breathing problems. The regulations require you and your employer to take care when dealing with asbestos:

* You should always assume that materials contain asbestos unless it is obvious that they do not.

* A record of the location and condition of asbestos should be kept.

* A risk assessment should be carried out if there is a chance that anyone will be exposed to asbestos.

The general advice is as follows:

* Do not remove the asbestos. It is not a hazard unless it is removed or damaged.

* Remember that not all asbestos presents the same risk. Asbestos cement is less dangerous than pipe insulation.

* Call in a specialist if you are uncertain.

Provision and Use of Work Equipment Regulations (PUWER) (1998)

PUWER concerns health and safety risks related to equipment used at work. It states that any risks arising from the use of equipment must either be prevented or controlled, and all suitable safety measures must have been taken. In addition, tools need to be:

* suitable for their intended use

* safe

REED TIP

Employers will want to know that you understand the importance of health and safety. Make sure you know the reasons for each safe working practice.

* well maintained

* used only by those who have been trained to do so.

Manual Handling Operations Regulations (1992)

These regulations try to control the risk of injury when lifting or handling bulky or heavy equipment and materials. The regulations state as follows:

* Hazardous manual handling should be avoided if possible.

* An assessment of hazardous manual handling should be made to try to find alternatives.

* You should use mechanical assistance where possible.

* The main idea is to look at how manual handling is carried out and finding safer ways of doing it.

Personal Protection at Work Regulations (PPE) (1992)

This law states that employers must provide employees with personal protective equipment (PPE) at work whenever there is a risk to health and safety. PPE needs to be:

* suitable for the work being done

* well maintained and replaced if damaged

* properly stored

* correctly used (which means employees need to be trained in how to use the PPE properly).

Work at Height Regulations (2005)

Whenever a person works at any height there is a risk that they could fall and injure themselves. The regulations place a duty on employers or anyone who controls the work of others. This means that they need to:

* plan and organise the work

* make sure those working at height are competent

* assess the risks and provide appropriate equipment

* manage work near or on fragile surfaces

* ensure equipment is inspected and maintained.

In all cases the regulations suggest that, if it is possible, work at height should be avoided. Perhaps the job could be done from ground level? If it is not possible, then equipment and other measures are needed to prevent the risk of falling. When working at height measures also need to be put in place to minimise the distance someone might fall.

KEY TERMS

PPE

– personal protective equipment can include gloves, goggles and hard hats.

Competent

– to be competent an organisation or individual must have:

* sufficient knowledge of the tasks to be undertaken and the risks involved

* the experience and ability to carry out their duties in relation to the project, to recognise their limitations and take appropriate action to prevent harm to those carrying out construction work, or those affected by the work.

(*Source* HSE)

Figure 1.1 Examples of personal protective equipment

Employer responsibilities under HASAWA

HASAWA states that employers with five or more staff need their own health and safety policy. Employers must assess any risks that may be involved in their workplace and then introduce controls to reduce these risks. These risk assessments need to be reviewed regularly.

Employers also need to supply personal protective equipment (PPE) to all employees when it is needed and to ensure that it is worn when required.

Specific employer responsibilities are outlined in Table 1.1.

Employee responsibilities under HASAWA

HASAWA states that all those operating in the workplace must aim to work in a safe way. For example, they must wear any PPE provided and look after their equipment. Employees should not be charged for PPE or any actions that the employer needs to take to ensure safety.

Specific employer responsibilities are outlined in Table 1.1. Table 1.2 identifies the key employee responsibilities.

KEY TERMS

Risk

– the likelihood that a person may be harmed if they are exposed to a hazard.

Hazard

– a potential source of harm, injury or ill-health.

Near miss

– any incident, accident or emergency that did not result in an injury but could have done so.

Employer responsibility	Explanation
Safe working environment	Where possible all potential risks and hazards should be eliminated.
Adequate staff training	When new employees begin a job their induction should cover health and safety. There should be ongoing training for existing employees on risks and control measures.
Health and safety information	Relevant information related to health and safety should be available for employees to read and have their own copies.
Risk assessment	Each task or job should be investigated and potential risks identified so that measures can be put in place. A risk assessment and method statement should be produced. The method statement will tell you how to carry out the task, what PPE to wear, equipment to use and the sequence of its use.
Supervision	A competent and experienced individual should always be available to help ensure that health and safety problems are avoided.

Table 1.1 Employer responsibilities under HASAWA

Employee responsibility	Explanation
Working safely	Employees should take care of themselves, only do work that they are competent to carry out and remove obvious hazards if they are seen.
Working in partnership with the employer	Co-operation is important and you should never interfere with or misuse any health and safety signs or equipment. You should always follow the site rules.
Reporting hazards, near misses and accidents correctly	Any health and safety problems should be reported and discussed, particularly a near miss or an actual accident.

Table 1.2 Employee responsibilities under HASAWA

Health and Safety Executive

The Health and Safety Executive (HSE) is responsible for health, safety and welfare. It carries out spot checks on different workplaces to make sure that the law is being followed.

HSE inspectors have access to all areas of a construction site and can also bring in the police. If they find a problem then they can issue an improvement notice. This gives the employer a limited amount of time to put things right.

In serious cases, the HSE can issue a prohibition notice. This means all work has to stop until the problem is dealt with. An employer, the employees or sub-contractors could be taken to court.

The roles and responsibilities of the HSE are outlined in Table 1.3.

Responsibility	Explanation
Enforcement	It is the HSE's responsibility to reduce work-related death, injury and ill health. It will use the law against those who put others at risk.
Legislation and advice	The HSE will use health and safety legislation to serve improvement or prohibition notices or even to prosecute those who break health and safety rules. Inspectors will provide advice either face-to-face or in writing on health and safety matters.
Inspection	The HSE will look at site conditions, standards and practices and inspect documents to make sure that businesses and individuals are complying with health and safety law.

Table 1.3 HSE roles and responsibilities

Sources of health and safety information

There is a wide variety of health and safety information. Most of it is available free of charge, while other organisations may make a charge to provide information and advice. Table 1.4 outlines the key sources of health and safety information.

Source	Types of information	Website
Health and Safety Executive (HSE)	The HSE is the primary source of work-related health and safety information. It covers all possible topics and industries.	www.hse.gov.uk
Construction Industry Training Board (CITB)	The national training organisation provides key information on legislation and site safety.	www.citb.co.uk
British Standards Institute (BSI)	Provides guidelines for risk management, PPE, fire hazards and many other health and safety-related areas.	www.bsigroup.com
Royal Society for the Prevention of Accidents (RoSPA)	Provides training, consultancy and advice on a wide range of health and safety issues that are aimed to reduce work related accidents and ill health.	www.rospa.com
Royal Society for Public Health (RSPH)	Has a range of qualifications and training programmes focusing on health and safety.	www.rsph.org.uk

Table 1.4 Health and safety information

Informing the HSE

The HSE requires the reporting of:

* deaths and injuries – any **major injury**, **over 7-day injury** or death
* occupational disease
* dangerous occurrence – a collapse, explosion, fire or collision
* gas accidents – any accidental leaks or other incident related to gas.

Enforcing guidance

Work-related injuries and illnesses affect huge numbers of people. According to the HSE, 1.1 million working people in the UK suffered from a work-related illness in 2011 to 2012. Across all industries, 173 workers were killed, 111,000 other injuries were reported and 27 million working days were lost.

The construction industry is a high risk one and, although only around 5 per cent of the working population is in construction, it accounts for 10 per cent of all major injuries and 22 per cent of fatal injuries.

The good news is that enforcing guidance on health and safety has driven down the numbers of injuries and deaths in the industry. Only 20 years ago over 120 construction workers died in workplace accidents each year. This is now reduced to fewer than 60 a year.

However, there is still more work to be done and it is vital that organisations such as the HSE continue to enforce health and safety and continue to reduce risks in the industry.

On-site safety inductions and toolbox talks

The HSE suggests that all new workers arriving on site should attend a short induction session on health and safety. It should:

* show the commitment of the company to health and safety
* explain the health and safety policy
* explain the roles individuals play in the policy
* state that each individual has a legal duty to contribute to safe working
* cover issues like excavations, work at height, electricity and fire risk
* provide a layout of the site and show evacuation routes
* identify where fire fighting equipment is located
* ensure that all employees have evidence of their skills
* stress the importance of signing in and out of the site.

Behaviour and actions that could affect others

It is the responsibility of everyone on site not only to look after their own health and safety, but also to ensure that their actions do not put anyone else at risk.

Trying to carry out work that you are not competent to do is not only dangerous to yourself but could compromise the safety of others.

Simple actions, such as ensuring that all of your rubbish and waste is properly disposed of, will go a long way to removing hazards on site that could affect others.

Just as you should not create a hazard, ignoring an obvious one is just as dangerous. You should always obey site rules and particularly the health and safety rules. You should follow any instructions you are given.

ACCIDENT AND EMERGENCY PROCEDURES

All sites will have specific procedures for dealing with accidents and emergencies. An emergency will often mean that the site needs to be evacuated, so you should know in advance where to assemble and who to report to. The site should never be re-entered without authorisation from an individual in charge or the emergency services.

Types of emergencies

Emergencies are incidents that require immediate action. They can include:

- fires
- spillages or leaks of chemicals or other hazardous substances, such as gas
- failure of a scaffold
- collapse of a wall or trench
- a health problem
- an injury
- bombs and security alerts.

Legislation and reporting accidents

RIDDOR (1995) puts a duty on employers, anyone who is self-employed, or an individual in control of the work, to report any serious workplace accidents, occupational diseases or dangerous occurrences (also known as near misses).

The report has to be made by these individuals and, if it is serious enough, the responsible person may have to fill out a RIDDOR report.

Figure 1.2 It's important that you know where your company's fire-fighting equipment is located

Injuries, diseases and dangerous occurrences

Construction sites can be dangerous places, as we have seen. The HSE maintains a list of all possible injuries, diseases and dangerous occurrences, particularly those that need to be reported.

Injuries

There are two main classifications of injuries: minor and major. A minor injury can usually be handled by a competent first aider, although it is often a good idea to refer the individual to their doctor or to the hospital. Typical minor injuries can include:

* minor cuts
* minor burns
* exposure to fumes.

Major injuries are more dangerous and will usually require the presence of an ambulance with paramedics. Major injuries can include:

* bone fracture
* concussion
* unconsciousness
* electric shock.

Diseases

There are several different diseases and health issues that have to be reported, particularly if a doctor notifies that a disease has been diagnosed. These include:

* poisoning
* infections
* skin diseases
* occupational cancer
* lung diseases
* hand/arm vibration syndrome.

Dangerous occurrences

Even if something happens that does not result in an injury, but could easily have done so, it is classed as a dangerous occurrence. It needs to be reported immediately and then followed up by an accident report form. Dangerous occurrences can include:

* accidental release of a substance that could damage health

* anything coming into contact with overhead power lines

* an electrical problem that caused a fire or explosion

* collapse or partial collapse of scaffolding over 5 m high.

PRACTICAL TIP

An up-to-date list of dangerous occurrences is maintained by the Health and Safety Executive.

Recording accidents and emergencies

The Reporting of Injuries, Diseases and Dangerous Occurrences Regulations (RIDDOR) (1995) requires employers to:

* report any relevant injuries, diseases or dangerous occurrences to the Health and Safety Executive (HSE)

* keep records of incidents in a formal and organised manner (for example, in an accident book or online database).

After an accident, you may need to complete an accident report form – either in writing or online. This form may be completed by the person who was injured or the first aider.

On the accident report form you need to note down:

* the casualty's personal details, e.g. name, address, occupation

* the name of the person filling in the report form

* the details of the accident.

In addition, the person reporting the accident will need to sign the form.

On site a trained first aider will be the first individual to try and deal with the situation. In addition to trying to save life, stop the condition from getting worse and getting help, they will also record the occurrence.

On larger sites there will be a safety officer, but all businesses should keep records and documentation that details any accident or emergency that has taken place under RIDDOR and to provide that information if the HSE requests it.

Importance of reporting accidents and near misses

Reporting incidents is not just about complying with the law or providing information for statistics. Each time an accident or near miss takes place it means lessons can be learned and future problems avoided.

The accident or near miss can alert the business or organisation to a potential problem. They can then take steps to ensure that it does not occur in the future.

Major and minor injuries and near misses

RIDDOR defines a major injury as:

* a fracture (but not to a finger, thumb or toes)

* a dislocation

* an amputation

* a loss of sight in an eye

* a chemical or hot metal burn to the eye

* a penetrating injury to the eye

* an electric shock or electric burn leading to unconsciousness and/or requiring resuscitation

* hyperthermia, heat-induced illness or unconsciousness

* asphyxia

* exposure to a harmful substance

* inhalation of a substance

* acute illness after exposure to toxins or infected materials.

A minor injury could be considered as any occurrence that does not fall into any of the above categories.

A near miss is any incident that did not actually result in an injury but which could have caused a major injury if it had done so. Non-reportable near misses are useful to record as they can help to identify potential problems. Looking at a list of near misses might show patterns for potential risk.

Accident trends

We have already seen that the HSE maintains statistics on the number and types of construction accidents. The following are among the 2011/2012 construction statistics:

* There were 49 fatalities.

* There were 5,000 occupational cancer patients.

* There were 74,000 cases of work-related ill health.

* The most common types of injury were caused by falls, although many injuries were caused by falling objects, collapses and electricity. A number of construction workers were also hurt when they slipped or tripped, or were injured while lifting heavy objects.

Accidents, emergencies and the employer

Even less serious accidents and injuries can cost a business a great deal of money. But there are other costs too:

* Poor company image – if a business does not have health and safety controls in place then it may get a reputation for not caring about its employees. The number of accidents and injuries may be far higher than average.

* Loss of production – the injured individual might have to be treated and then may need a period of time off work to recover. The loss of production can include those who have to take time out from working to help the injured person and the time of a manager or supervisor who has to deal with all the paperwork and problems.

* Insurance – each time there is an accident or injury claim against the company's insurance the premiums will go up. If there are many accidents and injuries the business may find it impossible to get insurance. It is a legal requirement for a business to have insurance so in the end that company might have to close down.

* Closure of site – if there is a serious accident or injury then the site may have to be closed while investigations take place to discover the reason, or who was responsible. This could cause serious delays and loss of income for workers and the business.

Accident and emergency authorised personnel

Several different groups of people could be involved in dealing with accident and emergency situations. These are listed in Table 1.5.

Authorised personnel	Role
First aiders and emergency responders	These are employees on site and in the workforce who have been trained to be the first to respond to accidents and injuries. The minimum provision of an appointed person would be someone who has had basic first aid training. The appointment of a first aider is someone who has attained a higher or specific level of training. A construction site with fewer than 5 employees needs an appointed first aider. A construction site with up to 50 employees requires a trained first aider, and for bigger sites at least one trained first aider is required for every 50 people.
Supervisors and managers	These have the responsibility of managing the site and would have to organise the response and contact emergency services if necessary. They would also ensure that records of any accidents are completed and up to date and notify the HSE if required.
Health and Safety Executive	The HSE requires businesses to investigate all accidents and emergencies. The HSE may send an inspector, or even a team, to investigate and take action if the law has been broken.
Emergency services	Calling the emergency services depends on the seriousness of the accident. Paramedics will take charge of the situation if there is a serious injury and if they feel it necessary will take the individual to hospital.

Table 1.5 People who deal with accident and emergency situations

DID YOU KNOW?

The three main emergency services in the UK are: the Fire Service (for fire and rescue); the Ambulance Service (for medical emergencies); the Police (for an immediate police response). Call them on 999 only if it is an emergency.

The basic first aid kit

BS 8599 relates to first aid kits, but it is not legally binding. The contents of a first aid box will depend on an employer's assessment of their likely needs. The HSE does not have to approve the contents of a first aid box but it states that where the work involves low level hazards the minimum contents of a first aid box should be:

* a copy of its leaflet on first aid – *HSE Basic advice on first aid at work*

* 20 sterile plasters of assorted size

* 2 sterile eye pads

* 4 sterile triangular bandages

* 6 safety pins

* 2 large sterile, unmedicated wound dressings

* 6 medium-sized sterile unmedicated wound dressings

* 1 pair of disposable gloves.

The HSE also recommends that no tablets or medicines are kept in the first aid box.

Figure 1.3 A typical first aid box

What to do if you discover an accident

When an accident happens it may not only injure the person involved directly, but it may also create a hazard that could then injure others. You need to make sure that the area is safe enough for you or someone else to help the injured person. It may be necessary to turn off the electrical supply or remove obstructions to the site of the accident.

The first thing that needs to be done if there is an accident is to raise the alarm. This could mean:

* calling for the first aider

* phoning for the emergency services

* dealing with the problem yourself.

How you respond will depend on the severity of the injury.

You should follow this procedure if you need to contact the emergency services:

* Find a telephone away from the emergency.

* Dial 999.

* You may have to go through a switchboard. Carefully listen to what the operator is saying to you and try to stay calm.

* When asked, give the operator your name and location, and the name of the emergency service or services you require.

* You will then be transferred to the appropriate emergency service, who will ask you questions about the accident and its location. Answer the questions in a clear and calm way.

* Once the call is over, make sure someone is available to help direct the emergency services to the location of the accident.

IDENTIFYING HAZARDS

As we have already seen, construction sites are potentially dangerous places. The most effective way of handling health and safety on a construction site is to spot the hazards and deal with them before they can cause an accident or an injury. This begins with basic housekeeping and carrying out risk assessments. It also means having a procedure in place to report hazards so that they can be dealt with.

Good housekeeping

Work areas should always be clean and tidy. Sites that are messy, strewn with materials, equipment, wires and other hazards can prove to be very dangerous. You should:

* always work in a tidy way

* never block fire exits or emergency escape routes

* never leave nails and screws scattered around

* ensure you clean and sweep up at the end of each working day

* not block walkways

* never overfill skips or bins

* never leave food waste on site.

Risk assessments and method statements

It is a legal requirement for employers to carry out risk assessments. This covers not only those who are actually working on a particular job, but other workers in the immediate area, and others who might be affected by the work.

It is important to remember that when you are carrying out work your actions may affect the safety of other people. It is important, therefore, to know whether there are any potential hazards. Once you know what these hazards are you can do something to either prevent or reduce them as a risk. Every job has potential hazards.

There are five simple steps to carrying out a risk assessment, which are shown in Table 1.6, using the example of repointing brickwork on the front face of a dwelling.

Step	Action	Example
1	Identify hazards	The property is on a street with a narrow pavement. The damaged brickwork and loose mortar need to be removed and placed in a skip below. Scaffolding has been erected. The road is not closed to traffic.
2	Identify who is at risk	The workers repointing are at risk as they are working at height. Pedestrians and vehicles passing are at risk from the positioning of the skip and the chance that debris could fall from height.
3	What is the risk from the hazard that may cause an accident?	The risk to the workers is relatively low as they have PPE and the scaffolding has been correctly erected. The risk to those passing by is higher, as they are unaware of the work being carried out above them.
4	Measures to be taken to reduce the risk	Station someone near the skip to direct pedestrians and vehicles away from the skip while the work is being carried out. Fix a secure barrier to the edge of the scaffolding to reduce the chance of debris falling down. Lower the bricks and mortar debris using a bucket or bag into the skip and not throwing them from the scaffolding. Consider carrying out the work when there are fewer pedestrians and less traffic on the road.
5	Monitor the risk	If there are problems with the first stages of the job, you need to take steps to solve them. If necessary consider taking the debris by hand through the building after removal.

Table 1.6 A five-step risk assessment for repointing brickwork

These working practices can help to prevent accidents or dangerous situations occurring in the workplace:

* *Risk assessments* look carefully at what could cause an individual harm and how to prevent this. This is to ensure that no one should be injured or become ill as a result of their work. Risk assessments identify how likely it is that an accident might happen and the consequences of it happening. A risk factor is worked out and control measures created to try to offset them.

* *Method statements,* however brief, should be available for every risk assessment. They summarise risk assessments and other findings to provide guidance on how the work should be carried out.

* *Permit to work systems* are used for very high risk or even potentially fatal activities. They are checklists that need to be completed before the work begins. They must be signed by a supervisor.

* *A hazard book* lists standard tasks and identifies common hazards. These are useful tools to help quickly identify hazards related to particular tasks.

Types of hazards

Typical construction accidents can include:

* fires and explosions
* slips, trips and falls.
* burns, including those from chemicals
* falls from scaffolding, ladders and roofs
* electrocution
* injury from faulty machinery
* power tool accidents
* being hit by construction debris
* falling through holes in flooring

We will look at some of the more common hazards in a little more detail.

Fires

Fires need oxygen, heat and fuel to burn. Even a spark can provide enough heat needed to start a fire, and anything flammable, such as petrol, paper or wood, provides the fuel. It may help to remember the 'triangle of fire' – heat, oxygen and fuel are all needed to make fire so remove one or more to help prevent or stop the fire.

Tripping

Leaving equipment and materials lying around can cause accidents, as can trailing cables and spilt water or oil. Some of these materials are also potential fire hazards.

Chemical spills

If the chemicals are not hazardous then they just need to be mopped up. But sometimes they do involve hazardous materials and there will be an existing plan on how to deal with them. A risk assessment will have been carried out.

Falls from height

A fall even from a low height can cause serious injuries. Precautions need to be taken when working at height to avoid permanent injury. You should also consider falls into open excavations as falls from height. All the same precautions need to be in place to prevent a fall.

Burns

Burns can be caused not only by fires and heat, but also from chemicals and solvents. Electricity and wet concrete and cement can also burn skin. PPE is often the best way to avoid these dangers. Sunburn is a common and uncomfortable form of burning and sunscreen should be made available. Keeping covered up, for example keeping skin covered up will help to prevent sunburn. You might think a tan looks good, but it could lead to skin cancer.

Electrical

Electricity is hazardous and electric shocks can cause burns and muscle damage, and can kill.

Exposure to hazardous substances

We look at hazardous substances in more detail on pages 20–1. COSHH regulations identify hazardous substances and require them to be labelled. You should always follow the instructions when using them.

Plant and vehicles

On busy sites there is always a danger from moving vehicles and heavy plant. Although many are fitted with reversing alarms, it may not be easy to hear them over other machinery and equipment. You should always ensure you are not blocking routes or exits. Designated walkways separate site traffic and pedestrians – this includes workers who are walking around the site. Crossing points should be in place for ease of movement on site.

Reporting hazards

We have already seen that hazards have the potential to cause serious accidents and injuries. It is therefore important to report hazards and there are different methods of doing this.

The first major reason to report hazards is to prevent danger to others, whether they are other employees or visitors to the site. It is vital to prevent accidents from taking place and to quickly correct any dangerous situations.

Injuries, diseases and actual accidents all need to be reported and so do dangerous occurrences. These are incidents that do not result in an actual injury, but could easily have hurt someone.

Accidents need to be recorded in an accident book, computer database or other secure recording system, as do near misses. Again it is a legal requirement to keep appropriate records of accidents and every company will have a procedure for this which they should tell you about. Everyone should know where the book is kept or how the records are made. Anyone that has been hurt or has taken part in dealing with an occurrence should complete the details of what has happened. Typically this will require you to fill in:

* the date, time and place of the incident

* how it happened

* what was the cause

* how it was dealt with

* who was involved

* signature and date.

The details in the book have to be transferred onto an official HSE report form.

As far as is possible, the site, company or workplace will have set procedures in place for reporting hazards and accidents. These procedures will usually be found in the place where the accident book or records are stored. The location tends to be posted on the site notice board.

How hazards are created

Construction sites are busy places. There are constantly new stages in development. As each stage is begun a whole new set of potential hazards need to be considered.

At the same time, new workers will always be joining the site. It is mandatory for them to be given health and safety instruction during induction. But sometimes this is impossible due to pressure of work or availability of trainers.

Construction sites can become even more hazardous in times of extreme weather:

* Flooding – long periods of rain can cause trenches to fill with water, cellars to be flooded and smooth surfaces to become extremely wet and slippery.

* Wind – strong winds may prevent all work at height. Scaffolding may have become unstable, unsecured roofing materials may come loose, dry-stored materials such as sand and cement may have been blown across the site.

* Heat – this can change the behaviour of materials: setting quicker, failing to cure and melting. It can also seriously affect the health of the workforce through dehydration and heat exhaustion.

* Snow – this can add enormous weight to roofs and other structures and could cause collapse. Snow can also prevent access or block exits and can mean that simple and routine work becomes impossible due to frozen conditions.

Storing combustibles and chemicals

A combustible substance can be both flammable and explosive. There are some basic suggestions from the HSE about storing these:

* Ventilation – the area should be well ventilated to disperse any vapours that could trigger off an explosion.

* Ignition – an ignition is any spark or flame that could trigger off the vapours, so materials should be stored away from any area that uses electrical equipment or any tool that heats up.

* Containment – the materials should always be kept in proper containers with lids and there should be spillage trays to prevent any leak seeping into other parts of the site.

* Exchange – in many cases it can be possible to find an alternative material that is less dangerous. This option should be taken if possible.

* Separation – always keep flammable substances away from general work areas. If possible they should be partitioned off.

Combustible materials can include a large number of commonly used substances, such as cleaning agents, paints and adhesives.

HEALTH AND HYGIENE

Just as hazards can be a major problem on site, other less obvious problems relating to health and hygiene can also be an issue. It is both your responsibility and that of your employer to make sure that you stay healthy.

The employer will need to provide basic welfare facilities, no matter where you are working and these must have minimum standards.

KEY TERMS

Contamination

– this is when a substance has been polluted by some harmful substance or chemical.

Welfare facilities

Welfare facilities can include a wide range of different considerations, as can be seen in Table 1.7.

Facilities	Purpose and minimum standards
Toilets	If there is a lock on the door there is no need to have separate male and female toilets. There should be enough for the site workforce. If there is no flushing water on site they must be chemical toilets.
Washing facilities	There should be a wash basin large enough to be able to wash up to the elbow. There should be soap, hot and cold water and, if you are working with dangerous substances, then showers are needed.
Drinking water	Clean drinking water should be available; either directly connected to the mains or bottled water. Employers must ensure that there is no contamination.
Dry room	This can operate also as a store room, which needs to be secure so that workers can leave their belongings there and also use it as a place to dry out if they have been working in wet weather, in which case a heater needs to be provided.
Work break area	This is a shelter out of the wind and rain, with a kettle, a microwave, tables and chairs. It should also have heating.

Table 1.7 Welfare facilities in the workplace

CASE STUDY

South Tyneside Council's Housing Company

Staying safe on site

Johnny McErlane finished his apprenticeship at South Tyneside Homes a year ago.

'I've been working on sheltered accommodation for the last year, so there are a lot of vulnerable and elderly people around. All the things I learnt at college from doing the health and safety exams comes into practice really, like taking care when using extension leads, wearing high-vis and correct footwear. It's not just about your health and safety, but looking out for others as well.

On the shelters, you can get a health and safety inspector who just comes around randomly, so you have to always be ready. It just becomes a habit once it's been drilled into you. You're health and safety conscious all the time.

The shelters also have a fire alarm drill every second Monday, so you've got to know the procedure involved there. When it comes to the more specialised skills, such as mouth-to-mouth and CPR, you might have a designated first aider on site who will have their skills refreshed regularly. Having a full first aid certificate would be valuable if you're working in construction.

You cover quite a bit of the first aid skills in college and you really have to know them because you're not always working on large sites. For example, you might be on the repairs team, working in people's houses where you wouldn't have a first aider, so you've got to have the basic knowledge yourself, just in case. All our vans have a basic first aid kit that's kept fully stocked.

The company keeps our knowledge current with these "toolbox talks", which are like refresher courses. They give you any new information that needs to be passed on to all the trades. It's a good way of keeping everyone up to date.'

Noise

Ear defenders are the best precaution to protect the ears from loud noises on site. Ear defenders are either basic ear plugs or ear muffs, which can be seen in Fig 1.13 on page 32.

The long-term impact of noise depends on the intensity and duration of the noise. Basically, the louder and longer the noise exposure, the more damage is caused. There are ways of dealing with this:

* Remove the source of the noise.

* Move the equipment away from those not directly working with it.

* Put the source of the noise into a soundproof area or cover it with soundproof material.

* Ask a supervisor if they can move all other employees away from that part of the site until the noise stops.

Substances hazardous to health

COSHH Regulations (see page 3) identify a wide variety of substances and materials that must be labelled in different ways.

Controlling the use of these substances is always difficult. Ideally, their use should be eliminated (stopped) or they should be replaced with something less harmful. Failing this, they should only be used in controlled or restricted areas. If none of this is possible then they should only be used in controlled situations.

If a hazardous situation occurs at work, then you should:

* ensure the area is made safe

* inform the supervisor, site manager, safety officer or other nominated person.

You will also need to report any potential hazards or near misses.

Personal hygiene

Construction sites can be dirty places to work. Some jobs will expose you to dust, chemicals or substances that can make contact with your skin or may stain your work clothing. It is good practice to wear suitable PPE as a first line of defence as chemicals can penetrate your skin. Whenever you have finished a job you should always wash your hands. This is certainly true before eating lunch or travelling home. It can be good practice to have dedicated work clothing, which should be washed regularly.

Always ensure you wash your hands and face and scrub your nails. This will prevent dirt, chemicals and other substances from contaminating your food and your home.

Make sure that you regularly wash your work clothing and either repair it or replace it if it becomes too worn or stained.

Health risks

The construction industry uses a wide variety of substances that could harm your health. You will also be carrying out work that could be a health risk to you, and you should always be aware that certain activities could cause long-term damage or even kill you if things go wrong. Unfortunately not all health risks are immediately obvious. It is important to make sure that from time to time you have health checks, particularly if you have been using hazardous substances. Table 1.8 outlines some potential health risks in a typical construction site.

KEY TERMS

Dermatitis

– this is an inflammation of the skin. The skin will become red and sore, particularly if you scratch the area. A GP should be consulted.

Leptospirosis

– this is also known as Weil's disease. It is spread by touching soil or water contaminated with the urine of wild animals infected with the leptospira bacteria. Symptoms are usually flu-like but in extreme cases it can cause organ failure.

Health risk	Potential future problems
Dust	The most dangerous potential dust is, of course, asbestos, which **should only be handled by specialists under controlled conditions**. But even brick dust and other fine particles can cause eye injuries, problems with breathing and even cancer.
Chemicals	Inhaling or swallowing dangerous chemicals could cause immediate, long-term damage to lungs and other internal organs. Skin problems include burns or skin can become very inflamed and sore. This is known as dermatitis.
Bacteria	Contact with waste water or soil could lead to a bacterial infection. The germs in the water or dirt could cause infection which will require treatment if they enter the body. The most extreme version is leptospirosis.
Heavy objects	Lifting heavy, bulky or awkward objects can lead to permanent back injuries that could require surgery. Heavy objects can also damage the muscles in all areas of the body.
Noise	Failure to wear ear defenders when you are exposed to loud noises can permanently affect your hearing. This could lead to deafness in the future.
Vibrating tools	Using machines that vibrate can cause a condition known as hand/arm vibration syndrome (HAVS) or vibration white finger, which is caused by injury to nerves and blood vessels. You will feel tingling that could lead to permanent numbness in the fingers and hands, as well as muscle weakness.
Cuts	Any open wound, no matter how small, leaves your body exposed to potential infections. Cuts should always be cleaned and covered, preferably with a waterproof dressing. The blood loss from deep cuts could make you feel faint and weak, which may be dangerous if you are working at height or operating machinery.
Sunlight	Most construction work involves working outside. There is a temptation to take advantage of hot weather and get a tan. But long-term exposure to sunshine means risking skin cancer so you should cover up and apply sun cream.
Head injuries	You should seek medical attention after any bump to the head. Severe head injuries could cause epilepsy, hearing problems, brain damage or death.

Table 1.8 Health risks in construction

HANDLING AND STORING MATERIALS AND EQUIPMENT

On a busy construction site it is often tempting not to even think about the potential dangers of handling equipment and materials. If something needs to be moved or collected you will just pick it up without any thought. It is also tempting just to drop your tools and other equipment when you have finished with them to deal with later. But abandoned equipment and tools can cause hazards both for you and for other people.

Safe lifting

Lifting or handling heavy or bulky items is a major cause of injuries on construction sites. So whenever you are dealing with a heavy load, it is important to carry out a basic risk assessment.

The first thing you need to do is to think about the job to be done and ask:

* Do I need to lift it manually or is there another way of getting the object to where I need it?

Consider any mechanical methods of transporting loads or picking up materials. If there really is no alternative, then ask yourself:

1. Do I need to bend or twist?
2. Does the object need to be lifted or put down from high up?
3. Does the object need to be carried a long way?
4. Does the object need to be pushed or pulled for a long distance?
5. Is the object likely to shift around while it is being moved?

If the answer to any of these questions is 'yes', you may need to adjust the way the task is done to make it safer.

Think about the object itself. Ask:

1. Is it just heavy or is it also bulky and an awkward shape?
2. How easy is it to get a good hand-hold on the object?
3. Is the object a single item or are there parts that might move around and shift the weight?
4. Is the object hot or does it have sharp edges?

Again, if you have answered 'yes' to any of these questions, then you need to take steps to address these issues.

It is also important to think about the working environment and where the lifting and carrying is taking place. Ask yourself:

1. Are the floors stable?

2. Are the surfaces slippery?

3. Will a lack of space restrict my movement?

4. Are there any steps or slopes?

5. What is the lighting like?

Before lifting and moving an object, think about the following:

* Check that your pathway is clear to where the load needs to be taken.

* Look at the product data sheet and assess the weight. If you think the object is too heavy or difficult to move then ask someone to help you. Alternatively, you may need to use a mechanical lifting device.

When you are ready to lift, gently raise the load. Take care to ensure the correct posture – you should have a straight back, with your elbows tucked in, your knees bent and your feet slightly apart.

Once you have picked up the load, move slowly towards your destination. When you get there, make sure that you do not drop the load but carefully place it down.

DID YOU KNOW?

Although many people regard the weight limit for lifting and/or moving heavy or awkward objects to be 20 kg, the HSE does not recommend safe weights. There are many things that will affect the ability of an individual to lift and carry particular objects and the risk that this creates, so manual handling should be avoided altogether where possible.

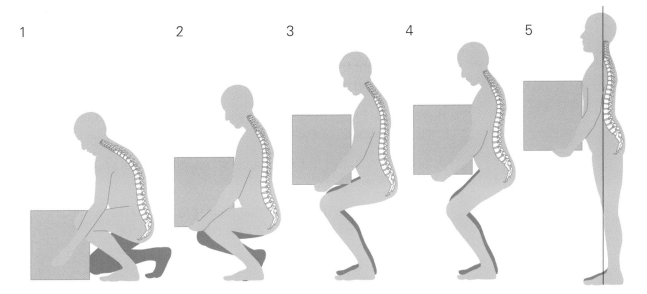

1   2   3   4   5

Figure 1.4 Take care to follow the correct procedure for lifting

Sack trolleys are useful for moving heavy and bulky items around. Gently slide the bottom of the sack trolley under the object and then raise the trolley to an angle of 45° before moving off. Make sure that the object is properly balanced and is not too big for the trolley.

Trailers and forklift trucks are often used on large construction sites, as are dump trucks. Never use these without proper training.

Figure 1.5 Pallet truck

Figure 1.6 Sack trolley

Site safety equipment

You should always read the construction site safety rules and when required wear your PPE. Simple things, such as wearing the right footwear for the right job, are important.

Safety equipment falls into two main categories:

* PPE – including hard hats, footwear, gloves, glasses and safety vests

* perimeter safety – this includes screens, netting and guards or clamps to prevent materials from falling or spreading.

Construction safety is also directed by signs, which will highlight potential hazards.

Safe handling of materials and equipment

All tools and equipment are potentially dangerous. It is up to you to make sure that they do not cause harm to yourself or others. You should always know how to use tools and equipment. This means either instruction from someone else who is experienced, or at least reading the manufacturer's instructions.

You should always make sure that you:

* use the right tool – don't be tempted to use a tool that is close to hand instead of the one that is right for the job

* wear your PPE – the one time you decide not to bother could be the time that you injure yourself

* never try to use a tool or a piece of equipment that you have not been trained to use.

You should always remember that if you are working on a building that was constructed before 2000 it may contain asbestos.

Correct storage

We have already seen that tools and equipment need to be treated with respect. Damaged tools and equipment are not only less effective at doing their job, they could also cause you to injure yourself.

Table 1.9 provides some pointers on how to store and handle different types of materials and equipment.

Materials and equipment	Safe storage and handling
Hand tools	Store hand tools with sharp edges either in a cover or a roll. They should be stored in bags or boxes. They should always be dried before putting them away as they will rust.
Power tools	Never carry them by the cable. Store them in their original carrying case. Always follow the manufacturer's instructions.
Wheelbarrows	Check the tyres and metal stays regularly. Always clean out after use and never overload.
Bricks and blocks	Never store more than two packs high. When cutting open a pack, be careful as the bricks could collapse.
Slabs and curbs	Store slabs flat on their edges on level ground, preferably with wood underneath to prevent damage. Store curbs the same way. To prevent weather damage, cover them with a sheet.
Tiles	Always cover them and protect them from damage as they are relatively fragile. Ideally store them in a hut or container.
Aggregates	Never store aggregates under trees as leaves will drop on them and contaminate them. Cover them with plastic sheets.
Plaster and plasterboard	Plaster needs to be kept dry, so even if stored inside you should take the precaution of putting the bags on pallets. To prevent moisture do not store against walls and do not pile higher than five bags. Plasterboard can be awkward to manage and move around. It also needs to be stored in a waterproof area. It should be stored flat and off the ground but should not be stored against walls as it may bend. Use a rotation system so that the materials are not stored in the same place for long periods.
Wood	Always keep wood in dry, well-ventilated conditions. If it needs to be stored outside it should be stored on bearers that may be on concrete. If wood gets wet and bends it is virtually useless. Always be careful when moving large cuts of wood or sheets of ply or MDF as they can easily become damaged.
Adhesives and paint	Always read the manufacturer's instructions. Ideally they should always be stored on clearly marked shelves. Make sure you rotate the stock using the older stock first. Always make sure that containers are tightly sealed. Storage areas must comply with fire regulations and display signs to advise of their contents.

Table 1.9 Safe storing and handling of materials and equipment

Waste control

The expectation within the building services industry is increasingly that working practices conserve energy and protect the environment. Everyone can play a part in this. For example, you can contribute by turning off hose pipes when you have finished using water, or not running electrical items when you don't need to.

Simple things, such as keeping construction sites neat and orderly, can go a long way to conserving energy and protecting the environment. A good way to remember this is Sort, Set, Shine, Standardise:

* Sort – sort and store items in your work area, eliminate clutter and manage deliveries.

* Set – everything should have its own place and be clearly marked and easy to access. In other words, be neat!

Figure 1.7 It's important to create as little waste as possible on the construction site

* Shine – clean your work area and you will be able to see potential problems far more easily.

* Standardise – by using standardised working practices you can keep organised, clean and safe.

Reducing waste is all about good working practice. By reducing wastage disposal, and recycling materials on site, you will benefit from savings on raw materials and lower transportation costs.

Planning ahead, and accurately measuring and cutting materials, means that you will be able to reduce wastage.

BASIC WORKING PLATFORMS AND ACCESS EQUIPMENT

Working at height should be eliminated or the work carried out using other methods where possible. However, there may be situations where you may need to work at height. These situations can include:

* roofing

* repair and maintenance above ground level

* working on high ceilings.

Any work at height must be carefully planned. Access equipment includes all types of ladder, scaffold and platform. You must always use a working platform that is safe. Sometimes a simple step ladder will be sufficient, but at other times you may have to use a tower scaffold.

Generally, ladders are fine for small, quick jobs of less than 30 minutes. However, for larger, longer jobs a more permanent piece of access equipment will be necessary.

Working platforms and access equipment: good practice and dangers of working at height

Table 1.10 outlines the common types of equipment used to allow you to work at heights, along with the basic safety checks necessary.

Equipment	Main features	Safety checks
Step ladder	Ideal for confined spaces. Four legs give stability	• Knee should remain below top of steps • Check hinges, cords or ropes • Position only to face work
Ladder	Ideal for basic access, short-term work. Made from aluminium, fibreglass or wood	• Check rungs, tie rods, repairs, and ropes and cords on stepladders • Ensure it is placed on firm, level ground • Angle should be no greater than 75° or 1 in 4
Mobile mini towers or scaffolds	These are usually aluminium and foldable, with lockable wheels	• Ensure the ground is even and the wheels are locked • Never move the platform while it has tools, equipment or people on it
Roof ladders and crawling boards	The roof ladder allows access while crawling boards provide a safe passage over tiles	• The ladder needs to be long enough and supported • Check boards are in good condition • Check the welds are intact • Ensure all clips function correctly
Mobile tower scaffolds	These larger versions of mini towers usually have edge protection	• Ensure the ground is even and the wheels are locked • Never move the platform while it has tools, equipment or people on it • Base width to height ratio should be no greater than 1:3
Fixed scaffolds and edge protection	Scaffolds fitted and sized to the specific job, with edge protection and guard rails	• There needs to be sufficient braces, guard rails and scaffold boards • The tubes should be level • There should be proper access using a ladder
Mobile elevated work platforms	Known as scissor lifts or cherry pickers	• Specialist training is required before use • Use guard rails and toe boards • Care needs to be taken to avoid overhead hazards such as cables

Table 1.10 Equipment for working at height and safety checks

You must be trained in the use of certain types of access equipment, like mobile scaffolds. Care needs to be taken when assembling and using access equipment. These are all examples of good practice:

Figure 1.8 A tower scaffold

● Step ladders should always rest firmly on the ground. Only use the top step if the ladder is part of a platform.

● Do not rest ladders against fragile surfaces, and always use both hands to climb. It is best if the ladder is steadied (footed) by someone at the foot of the ladder. Always maintain three points of contact – two feet and one hand.

● A roof ladder is positioned by turning it on its wheels and pushing it up the roof. It then hooks over the ridge tiles. Ensure that the access ladder to the roof is directly beside the roof ladder.

● A mobile scaffold is put together by slotting sections until the required height is reached. The working platform needs to have a suitable edge protection such as guard-rails and toe-boards. Always push from the bottom of the base and not from the top to move it, otherwise it may lean or topple over.

WORKING SAFELY WITH ELECTRICITY

It is essential whenever you work with electricity that you are competent and that you understand the common dangers. Electrical tools must be used in a safe manner on site. There are precautions that you can take to prevent possible injury, or even death.

Precautions

Whether you are using electrical tools or equipment on site, you should always remember the following:

* Use the right tool for the job.

* Use a transformer with equipment that runs on 110V.

* Keep the two voltages separate from each other. You should avoid using 230V where possible but use a residual current device (RCD) if you have to use 230V.

* When using 100V, ensure that leads are yellow in colour.

* Check the plug is in good order

* Confirm that the fuse is the correct rating for the equipment.

* Check the cable (including making sure that it does not present a tripping hazard).

* Find out where the mains switch is, in case you need to turn off the power in the event of an emergency.

* Never attempt to repair electrical equipment yourself.

* Disconnect from the mains power before making adjustments, such as changing a drill bit.

* Make sure that the electrical equipment has a sticker that displays a recent test date.

Visual inspection and testing is a three-stage process:

1. The user should check for potential danger signs, such as a frayed cable or cracked plug.

2. A formal visual inspection should then take place. If this is done correctly then most faults can be detected.

3. Combined inspections and PAT should take place at regular intervals by a competent person.

Watch out for the following causes of accidents – they would also fail a safety check:

KEY TERMS

PAT

– Portable Appliance Testing – regular testing is a health and safety requirement under the Electricity at Work Regulations (1989).

* damage to the power cable or plug

* taped joints on the cable

* wet or rusty tools and equipment

* weak external casing

* loose parts or screws

* signs of overheating

* the incorrect fuse

* lack of cord grip

* electrical wires attached to incorrect terminals

* bare wires.

When preparing to work on an electrical circuit, do not start until a permit to work has been issued by a supervisor or manager to a competent person.

Make sure the circuit is broken before you begin. A 'dead' circuit will not cause you, or anybody else, harm. These steps must be followed:

* Switch off – ensure the supply to the circuit is switched off by disconnecting the supply cables or using an isolating switch.

* Isolate – disconnect the power cables or use an isolating switch.

* Warn others – to avoid someone reconnecting the circuit, place warning signs at the isolation point.

* Lock off – this step physically prevents others from reconnecting the circuit.

* Testing – is carried out by electricians but you should be aware that it involves three parts:

    1. testing a voltmeter on a known good source (a live circuit) so you know it is working properly

    2. checking that the circuit to be worked on is dead

    3. rechecking your voltmeter on the known live source, to prove that it is still working properly.

It is important to make sure that the correct point of isolation is identified. Isolation can be next to a local isolation device, such as a plug or socket, or a circuit breaker or fuse.

The isolation should be locked off using a unique key or combination. This will prevent access to a main isolator until the work has been completed. Alternatively, the handle can be made detachable in the OFF position so that it can be physically removed once the circuit is switched off.

Dangers

You are likely to encounter a number of potential dangers when working with electricity on construction sites or in private houses. Table 1.11 outlines the most common dangers.

Danger	Identifying the danger
Faulty electrical equipment	Visually inspect for signs of damage. Equipment should be double insulated or incorporate an earth cable.
Damaged or worn cables	Check for signs of wear or damage regularly. This includes checking power tools and any wiring in the property.
Trailing cables	Cables lying on the ground, or worse, stretched too far, can present a tripping hazard. They could also be cut or damaged easily.
Cables and pipe work	Always treat services you find as though they are live. This is very important as services can be mistaken for one another. You may have been trained to use a cable and pipe locator that finds cables and metal pipes.
Buried or hidden cables	Make sure you have plans. Alternatively, use a cable and pipe locator, mark the positions, look out for signs of service connection cables or pipes and hand-dig trial holes to confirm positions.
Inadequate over-current protection	Check circuit breakers and fuses are the correct size current rating for the circuit. A qualified electrician may have to identify and label these.

Table 1.11 Common dangers when working with electricity

Each year there are around 1,000 accidents at work involving electric shocks or burns from electricity. If you are working in a construction site you are part of a group that is most at risk. Electrical accidents happen when you are working close to equipment that you think is disconnected but which is, in fact, live.

Another major danger is when electrical equipment is either misused or is faulty. Electricity can cause fires and contact with the live parts can give you an electric shock or burn you.

Different voltages

The two most common voltages that are used in the UK are 230V and 110V:

* 230V: this is the standard domestic voltage. But on construction sites it is considered to be unsafe and therefore 110V is commonly used.

* 110V: these plugs are marked with a yellow casement and they have a different shaped plug. A transformer is required to convert 230V to 110V.

Some larger homes, as well as industrial and commercial buildings, may have 415V supplies. This is the same voltage that is found on overhead electricity cables. In most houses and other buildings the voltage from these cables is reduced to 230V. This is what most electrical equipment works from. Some larger machinery actually needs 415V.

In these buildings the 415V comes into the building and then can either be used directly or it is reduced so that normal 230V appliances can be used.

Colour coded cables

Normally you will come across three differently coloured wires: Live, Neutral and Earth. These have standard colours that comply with European safety standards and to ensure that they are easily identifiable. However, in some older buildings the colours are different.

Wire type	Modern colour	Older colour
Live	Brown	Red
Neutral	Blue	Black
Earth	Yellow and Green	Yellow and Green

Table 1.12 Colour coding of cables

Working with equipment with different electrical voltages

You should always check that the electrical equipment that you are going to use is suitable for the available electrical supply. The equipment's power requirements are shown on its rating plate. The voltage from the supply needs to match the voltage that is required by the equipment.

Storing electrical equipment

Electrical equipment should be stored in dry and secure conditions. Electrical equipment should never get wet but – if it does happen – it should be dried before storage. You should always clean and adjust the equipment before connecting it to the electricity supply.

PERSONAL PROTECTIVE EQUIPMENT (PPE)

Personal protective equipment, or PPE, is a general term that is used to describe a variety of different types of clothing and equipment that aim to help protect against injuries or accidents. Some PPE you will use on a daily basis and others you may use from time to time. The type of PPE you wear depends on what you are doing and where you are. For example, the practical exercises in this book were photographed at a college, which has rules and requirements for PPE that are different to those on large construction sites. Follow your tutor's or employer's instructions at all times.

Types of PPE

PPE literally covers from head to foot. Here are the main PPE types.

Figure 1.9 A hi-vis jacket

Figure 1.10 Safety glasses and goggles

Figure 1.11 Hand protection

Figure 1.12 Head protection

Figure 1.13 Hearing protection

Protective clothing

Clothing protection such as overalls:

* provides some protection from spills, dust and irritants
* can help protect you from minor cuts and abrasions
* reduces wear to work clothing underneath.

Sometimes you may need waterproof or chemical-resistant overalls.

High visibility (hi-vis) clothing stands out against any background or in any weather conditions. It is important to wear high visibility clothing on a construction site to ensure that people can see you easily. In addition, workers should always try to wear light-coloured clothing underneath, as it is easier to see.

You need to keep your high visibility and protective clothing clean and in good condition.

Employers need to make sure that employees understand the reasons for wearing high visibility clothing and the consequences of not doing so.

Eye protection

For many jobs, it is essential to wear goggles or safety glasses to prevent small objects, such as dust, wood or metal, from getting into the eyes. As goggles tend to steam up, particularly if they are being worn with a mask, safety glasses can often be a good alternative.

Hand protection

Wearing gloves will help to prevent damage or injury to the hands or fingers. For example, general purpose gloves can prevent cuts, and rubber gloves can prevent skin irritation and inflammation, such as contact dermatitis caused by handling hazardous substances. There are many different types of gloves available, including specialist gloves for working with chemicals.

Head protection

Hard hats or safety helmets are compulsory on building sites. They can protect you from falling objects or banging your head. They need to fit well and they should be regularly inspected and checked for cracks. Worn straps mean that the helmet should be replaced, as a blow to the head can be fatal. Hard hats bear a date of manufacture and should be replaced after about 3 years.

Hearing protection

Ear defenders, such as ear protectors or plugs, aim to prevent damage to your hearing or hearing loss when you are working with loud tools or are involved in a very noisy job.

Respiratory protection

Breathing in fibre, dust or some gases could damage the lungs. Dust is a very common danger, so a dust mask, face mask or respirator may be necessary.

Make sure you have the right mask for the job. It needs to fit properly otherwise it will not give you sufficient protection.

Foot protection

Foot protection is compulsory on site. Footwear should include steel toecaps (or equivalent) to protect feet from dropped objects, midsole protection (usually a steel plate) to protect against puncture or penetration from things like nails on the floor, and soles with good grip to help prevent slips on wet surfaces.

Legislation covering PPE

The most important piece of legislation is the Personal Protective Equipment at Work Regulations (1992). It covers all sorts of PPE and sets out your responsibilities and those of the employer. Linked to this are the Control of Substances Hazardous to Health (2002) and the Provision and Use of Work Equipment Regulations (1992 and 1998).

Figure 1.14 Respiratory protection

Storing and maintaining PPE

All forms of PPE will be less effective if they are not properly maintained. This may mean examining the PPE and either replacing or cleaning it, or if relevant testing or repairing it. PPE needs to be stored properly so that it is not damaged, contaminated or lost. Each type of PPE should have a CE mark. This shows that it has met the necessary safety requirements.

Importance of PPE

PPE needs to be suitable for its intended use and it needs to be used in the correct way. As a worker or an employee you need to:

* make sure you are trained to use PPE

* follow your employer's instructions when using the PPE and always wear it when you are told to do so

* look after the PPE and if there is a problem with it report it.

Your employer will:

* know the risks that the PPE will either reduce or avoid

* know how the PPE should be maintained

* know its limitations.

Consequences of not using PPE

The consequences of not using PPE can be immediate or long-term. Immediate problems are more obvious, as you may injure yourself. The longer-term consequences could be ill health in the future. If your employer has provided PPE, you have a legal responsibility to wear it.

FIRE AND EMERGENCY PROCEDURES

KEY TERMS

Assembly point

– an agreed place outside the building to go to if there is an emergency.

If there is a fire or an emergency, it is vital that you raise the alarm quickly. You should leave the building or site and then head for the **assembly point.**

When there is an emergency a general alarm should sound. If you are working on a larger and more complex construction site, evacuation may begin by evacuating the area closest to the emergency. Areas will then be evacuated one-by-one to avoid congestion of the escape routes.

Figure 1.15 Assembly point sign

Three elements essential to creating a fire

Three ingredients are needed to make something combust (burn):

* oxygen   * heat   * fuel.

The fuel can be anything which burns, such as wood, paper or flammable liquids or gases, and oxygen is in the air around us, so all that is needed is sufficient heat to start a fire.

The fire triangle represents these three elements visually. By removing one of the three elements the fire can be prevented or extinguished.

Figure 1.16 The fire triangle

How fire is spread

Fire can easily move from one area to another by finding more fuel. You need to consider this when you are storing or using materials on site, and be aware that untidiness can be a fire risk. For example, if there are wood shavings on the ground the fire can move across them, burning up the shavings.

Heat can also transfer from one source of fuel to another. If a piece of wood is on fire and is against or close to another piece of wood, that too will catch fire and the fire will have spread.

On site, fires are classified according to the type of material that is on fire. This will determine the type of fire-fighting equipment you will need to use. The five different types of fire are shown in Table 1.13.

Class of fire	Fuel or material on fire
A	Wood, paper and textiles
B	Petrol, oil and other flammable liquids
C	LPG, propane and other flammable gases
D	Metals and metal powder
E	Electrical equipment

Table 1.13 Different classes of fire

There is also F, cooking oil, but this is less likely to be found on site, except in a kitchen.

Taking action if you discover a fire and fire evacuation procedures

During induction, you will have been shown what to do in the event of a fire and told about assembly points. These are marked by signs and somewhere on the site there will be a map showing their location.

If you discover a fire you should:

* sound the alarm

* not attempt to fight the fire unless you have had fire marshal training

* otherwise stop work, do not collect your belongings, do not run, and do not re-enter the site until the all clear has been given.

Different types of fire extinguishers

Extinguishers can be effective when tackling small localised fires. However, you must use the correct type of extinguisher. For example, putting water on an oil fire could make it explode. For this reason, you should not attempt to use a fire extinguisher unless you have had proper training.

When using an extinguisher it is important to remember the following safety points:

* Only use an extinguisher at the early stages of a fire, when it is small.

* The instructions for use appear on the extinguisher.

* If you do choose to fight the fire because it is small enough, and you are sure you know what is burning, position yourself between the fire and the exit, so that if it doesn't work you can still get out.

Type of fire risk	Fire class Symbol	White label Water	Cream label Foam	Black label Carbon dioxide	Blue label Dry powder	Yellow label Wet chemical
A – Solid (e.g. wood or paper)	A	✓	✓	✗	✓	✓
B – Liquid (e.g. petrol)	B	✗	✓	✓	✓	✗
C – Gas (e.g. propane)	C	✗	✗	✓	✓	✗
D – Metal (e.g. aluminium)	D METAL	✗	✗	✗	✓	✗
E – Electrical (i.e. any electrical equipment)	E	✗	✗	✓	✓	✗
F – Cooking oil (e.g. a chip pan)	F	✗	✗	✗	✗	✓

Table 1.14 Types of fire extinguishers

There are some differences you should be aware of when using different types of extinguisher:

* CO$_2$ extinguishers – do not touch the nozzle; simply operate by holding the handle. This is because the nozzle gets extremely cold when ejecting the CO$_2$, as does the canister. Fires put out with a CO$_2$ extinguisher may reignite, and you will need to ventilate the room after use.

* Powder extinguishers – these can be used on lots of kinds of fire, but can seriously reduce visibility by throwing powder into the air as well as on the fire.

SIGNS AND SAFETY NOTICES

In a well-organised working environment safety signs will warn you of potential dangers and tell you what to do to stay safe. They are used to warn you of hazards. Their purpose is to prevent accidents. Some will tell you what to do (or not to do) in particular parts of the site and some will show you where things are, such as the location of a first aid box or a fire exit.

Types of signs and safety notices

There are five basic types of safety sign, as well as signs that are a combination of two or more of these types. These are shown in Table 1.15.

Type of safety sign	What it tells you	What it looks like	Example
Prohibition sign	Tells you what you must *not* do	Usually round, in red and white	Do not use ladder
Hazard sign	Warns you about hazards	Triangular, in yellow and black	Caution Slippery floor
Mandatory sign	Tells you what you *must* do	Round, usually blue and white	Masks must be worn in this area
Safe condition or information sign	Gives important information, e.g. about where to find fire exits, assembly points or first aid kit, or about safe working practices	Green and white	First aid
Firefighting sign	Gives information about extinguishers, hydrants, hoses and fire alarm call points, etc.	Red with white lettering	Fire alarm call point
Combination sign	These have two or more of the elements of the other types of sign, e.g. hazard, prohibition and mandatory		DANGER Isolate before removing cover

Table 1.15 Different types of safety signs

TEST YOURSELF

1. Which of the following requires you to tell the HSE about any injuries or diseases?

   a. HASAWA

   b. COSHH

   c. RIDDOR

   d. PUWER

2. What is a prohibition notice?

   a. An instruction from the HSE to stop all work until a problem is dealt with

   b. A manufacturer's announcement to stop all work using faulty equipment

   c. A site contractor's decision not to use particular materials

   d. A local authority banning the use of a particular type of brick

3. Which of the following is considered a major injury?

   a. Bruising on the knee

   b. Cut

   c. Concussion

   d. Exposure to fumes

4. If there is an accident on a site who is likely to be the first to respond?

   a. First aider

   b. Police

   c. Paramedics

   d. HSE

5. Which of the following is a summary of risk assessments and is used for high risk activities?

   a. Permit to work

   b. Hazard book

   c. Monitoring statement

   d. Method statement

6. Some substances are combustible. Which of the following are examples of combustible materials?

   a. Adhesives

   b. Paints

   c. Cleaning agents

   d. All of these

7. What is dermatitis?

   a. Inflammation of the skin

   b. Inflammation of the ear

   c. Inflammation of the eye

   d. Inflammation of the nose

8. Screens, netting and guards on a site are all examples of which of the following?

   a. PPE

   b. Signs

   c. Perimeter safety

   d. Electrical equipment

9. Which of the following are also known as scissor lifts or cherry pickers?

   a. Bench saws

   b. Hand-held power tools

   c. Cement additives

   d. Mobile elevated work platforms

10. In older properties the neutral electricity wire is which colour?

    a. Black

    b. Red

    c. Blue

    d. Brown

Chapter 2

Unit CSA–L1Core02

KNOWLEDGE OF TECHNICAL INFORMATION, QUANTITIES AND COMMUNICATION WITH OTHERS

LEARNING OUTCOMES

LO1: Know how to interpret construction related technical information

LO2: Know how to determine quantities of materials

LO3: Know how to relay information in the construction environment

LO4: Know how to communicate with others in the construction environment

INTRODUCTION

The aim of this chapter is to:

* show you the processes of passing on information
* show you the concepts of effective communication.

INTERPRETING CONSTRUCTION-RELATED TECHNICAL INFORMATION

Even quite simple construction projects will require documents. These provide you with the necessary information you will need to do the job. The documents are produced by a range of different people and each document has a different purpose. Together they give you the full picture of the job, from the basic outline through to the technical specifications.

Importance of documentation

In many industries a great deal of information is only ever stored electronically. This is not always an option in the construction industry. Many documents, such as working drawings, will need to be referred to on site. Detailed drawings of components that need to be made will have to be measured and checked before making joints, for example, in the workshop.

It is not always easy to store and look after working documents. The following advice is worth remembering:

* Always ensure you have the latest version of a document to work from before you begin to follow its instructions.

* If you are not going to need to use a document until later then get into the habit of storing it somewhere safe.

* Try to make sure that you do not leave documents lying around on site, where they could get lost or damaged.

* Try to make sure you always have a second copy of the document. You should keep this away from the site, in reserve, in case you lose your working copy.

* You should store any documents that you have used on a particular job at least until that job is completely finished.

* You might need to store the documents for some time after in case you need to refer back to them for repair and servicing.

Interpreting construction specifications

Obviously it would be impossible to put in all of the details in full, so symbols, hatchings and abbreviations are used to simplify the drawings. All of these symbols or hatchings are drawn to follow a British Standards-approved format, BS 1192. The symbols cover various types of brickwork and blockwork, as well as concrete, hard core and insulation, as can be seen in Fig 2.1.

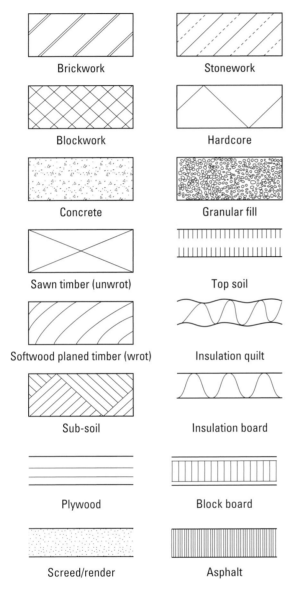

Figure 2.1 Symbols used on drawings

Common abbreviations

For the same reason, abbreviations are often used. Table 2.1 outlines some examples that you will need to become familiar with.

Abbreviation	Meaning
bwk	Brickwork
conc	Areas that will be concreted
dpc	Damp-proof course
fdn	Foundations
insul	Insulation
rwg	Rainwater gulleys
svp	Soil and vent pipe

Table 2.1

Types of documentation

Supporting information can be found in a variety of different types of documents. These include:

* drawings and plans
* programmes of work
* procedures
* specifications
* policies
* schedules
* manufacturers' technical information
* organisational documentation
* training and development records
* risk and method statements
* Construction (Design and Management) (CDM) Regulations
* Building Regulations.

Drawings and plans

Drawings are an important part of construction work. You will need to understand how drawings provide you with the information you need to carry out the work. The drawings show what the building will look like and how it will be constructed. This means that there are several different drawings of the building from different viewpoints.

Block plans

Block plans show the construction site and the surrounding area. Normally block plans are at a ratio of 1:2500 (usually in rural areas) and 1:1250 (usually in urban areas). This means that 1 mm on a block plan is equal to 2,500 mm or 1,250 mm on the ground.

Site plan

The site plan drawing shows what is basically planned for the site. It is an important drawing because it has been created in order to get approval for the project from planning committees or funding sources. In most cases the site plan is an architectural plan, showing the basic arrangement of buildings and any landscaping.

The site plan will usually show:

* directional orientation (i.e. the north point)

* location and size of the building or buildings

* existing structures

* clear measurements.

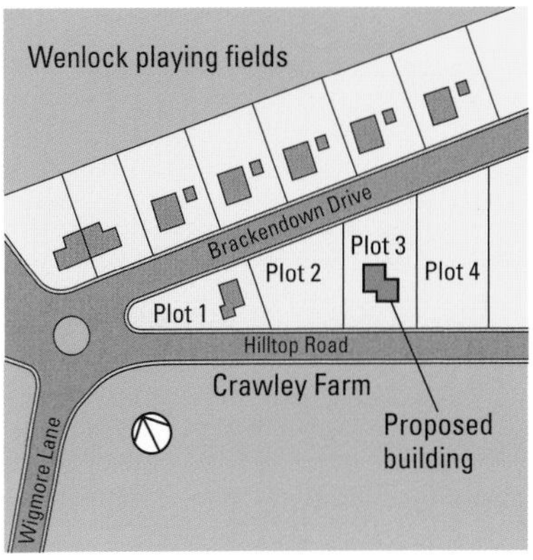

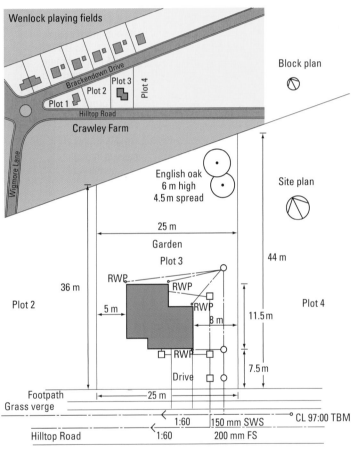

Figure 2.2 Block plan

General location

Location drawings show the site or building in relation to its surroundings. It will therefore show details such as boundaries, other buildings and roads. It will also contain other vital information, including:

* access   * sewers

* drainage   * the north point.

The scale will also be shown and the drawing will have a title. It will also be given a job or project number to help identify it easily, as well as an address, the date of the drawing and the name of the client. A version number will also be on the drawing with an amendment date if there have been any changes. You'll need to make sure you have the latest drawing.

Normally location drawings are either 1:500 or 1:200 (that is, 1 mm of the drawing represents 500 mm or 200 mm on the ground).

Figure 2.3 Location plan

Assembly

These are detailed drawings that illustrate the different elements and components of the construction. They tend to be 1:20, 1:10 or 1:5 (1 cm of the drawing represents 20 mm, 10 mm or 5 mm on the ground). This larger scale allows more detail to be shown, to ensure accurate construction.

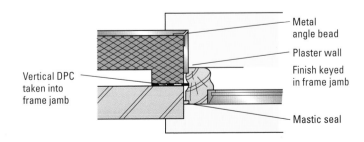

Figure 2.4 Assembly drawing

43

Sectional

These drawings aim to provide:

* vertical dimensions

* horizontal dimensions

* constructional details.

They can be used to show the height of ground levels, damp-proof courses, foundations and other aspects of the construction.

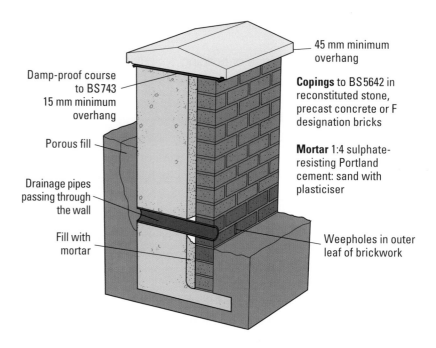

45 mm minimum overhang

Damp-proof course to BS743 15 mm minimum overhang

Copings to BS5642 in reconstituted stone, precast concrete or F designation bricks

Porous fill

Mortar 1:4 sulphate-resisting Portland cement: sand with plasticiser

Drainage pipes passing through the wall

Fill with mortar

Weepholes in outer leaf of brickwork

Figure 2.5 Section drawing of an earth retaining wall

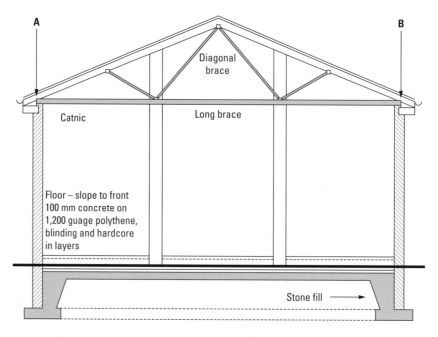

A

B

Diagonal brace

Catnic

Long brace

Floor – slope to front 100 mm concrete on 1,200 guage polythene, blinding and hardcore in layers

Stone fill

Figure 2.6 Section drawing of a garage

Detail drawings

These drawings show how a component needs to be manufactured. They are used to show the relationship between different components within the fabric of the building. For instance, an eaves detail would show rafters, wall plate, roof coverings, inner and outer masonry, insulation and much more. Details can be shown in various scales, but mainly 1:10, 1:5 and 1:1 (the same size as the actual component if it is small).

Orthographic projection (first angle)

First angle projection is a view that represents the side view, the front view and the plan view from above, as can be seen in Fig 2.8.

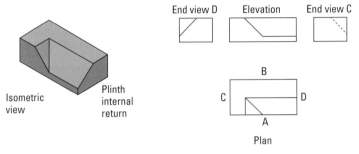

Figure 2.7 First angle projection

Serving hatch   Vertical section

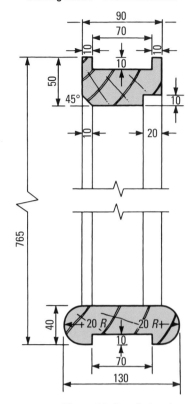

Figure 2.8 Detail drawing (measurements in mm)

Isometric projection

Isometric projection is a way of representing three dimensional objects in two dimensions, as can also be seen in Fig 2.8. All horizontal lines are drawn at 30°.

Programmes of work

Programmes of work show the actual sequence of any work activities on a construction project. Part of the work programme plan is to show target times. They are usually shown in the form of a bar or Gantt chart (a special kind of bar chart), as can be seen in Fig 2.9.

DID YOU KNOW?

First angle is also known as European projection because Americans use third angle projection, which shows the views from a different position.

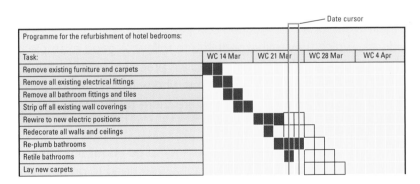

Figure 2.9 Single line contract plan Gantt chart

This figure shows the following:

* On the left hand side all of the tasks are listed – note this is ordered in the sequence of construction.

* On the right the blocks show the target start and end date for each of the individual tasks.

* The timescale can be either days, weeks or months.

Far more complex forms of work programmes can also be created. Fig 2.10 shows the construction of a house.

This more complex example of a Gantt chart shows the following:

* There are two lines – they show the target dates and actual dates. The actual dates are shaded, showing when the work actually began and how long it actually took.

* If this Gantt chart is kept up to date an accurate picture of progress and estimated completion time can be seen.

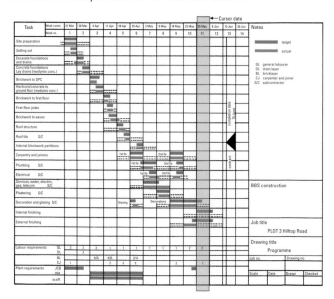

Figure 2.10 Gantt chart for the construction of a house

Procedures

When you work for a construction company there will be a series of procedures, which they will expect you to follow. A good example is the emergency procedure. This will explain precisely what is required in the case of an emergency on site and who will have responsibility to carry out particular duties. Procedures are there to show you the right way of doing something.

A construction procedure could outline how to go about building a wall or hanging a door, taking into account what you need to do beforehand, the materials and tools that are required, and the order in which you must carry out each step.

Another good example of a procedure is the procurement or buying procedure. This will outline:

* who is authorised to buy what, and how much individuals are allowed to spend

* any forms or documents that have to be completed when buying.

Specifications

In addition to drawings it is usually necessary to have documents known as specifications. These provide much more information, as can be seen in Fig 2.11.

The specifications give you a precise description. They will include:

* the address and description of the site

* on-site services (e.g. water and electricity)

* materials description, outlining the size, finish, quality and tolerances

* specific requirements, such as the individual who will authorise or approve work carried out

* any restrictions on site, such as working hours.

Policies

Policies are sets of principles or a programme of actions. The following are two good examples:

* The environmental policy outlines how the business goes about protecting the environment.

* The safety policy outlines how the business deals with health and safety matters and who is responsible for monitoring and maintaining it.

You will normally find both policies and procedures in site rules. These are usually explained to each new employee when they first join the company. Sometimes there may be additional site rules, depending on the job and the location of the work.

Schedules

Schedules are cross-referenced to drawings that have been prepared by an architect. They will show specific design information. Usually they are prepared for jobs that will be carried out regularly on site, such as:

* working on windows, doors, floors, walls or ceilings

* working on drainage, lintels or sanitary ware.

A schedule can be seen in Fig 2.12.

The schedule is very useful for:

* working out the quantities of materials needed

* ordering materials and components and then checking them against deliveries

* locating where specific materials will be used.

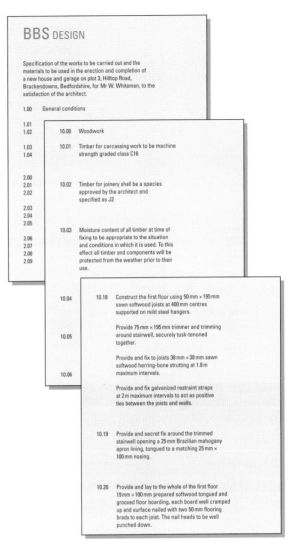

Figure 2.11 Extracts from a typical specification

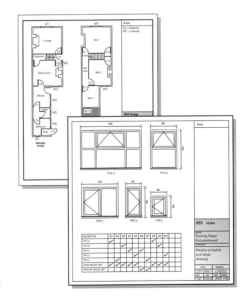

Figure 2.12 Typical windows schedule, range drawing and floor plans

Manufacturers' technical information

Almost everything that is bought to be used on site will come with a variety of information. The basic technical information provided will show what the equipment or material is intended to be used for, how it should be stored and any particular requirements it may have, such as handling or maintenance.

Technical information from the manufacturer can come from a variety of different sources:

* printed or downloadable data sheets

* printed or downloadable user instructions

* manufacturers' catalogues or brochures

* manufacturers' websites.

Organisational documentation

There is a huge potential list of organisational documentation and paperwork. Examples are outlined in Table 2.2. Visual examples can be seen in Fig 2.13 to 2.17.

Document	Purpose
Timesheet	Record of hours that you have worked and the jobs that you have carried out. This is used to help work out your wages and the total cost of the job.
Day worksheet	These detail work that has been carried out without providing an estimate beforehand. They usually include repairs or extra work and alterations.
Variation order	These are provided by the architect and given to the builder, showing any alterations, additions or omissions to the original job.
Confirmation notice	Provided by the architect to confirm any verbal instructions.
Daily report or site diary	These record things that might affect the project like detailed weather conditions, late deliveries or site visitors.
Orders and requisitions	These are order forms, requesting the delivery of materials.
Delivery notes	These are provided by the supplier of materials as a list of all materials being delivered. These need to be checked against materials actually delivered. The buyer will sign the delivery note when they are happy with the delivery.
Delivery record	These are lists of all materials that have been delivered on site.
Memorandum	These are used for internal communications and are usually brief.
Letters	These are used for external communications, usually to customers or suppliers.
Fax	Even though email is commonly used, the industry is still in favour of using faxes, as they provide an exact copy of an original document.

Table 2.2

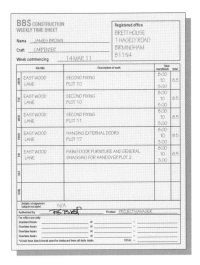

Figure 2.13 Timesheet

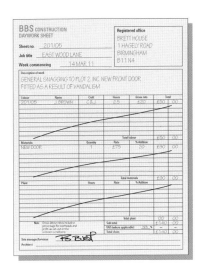

Figure 2.14 Day worksheet

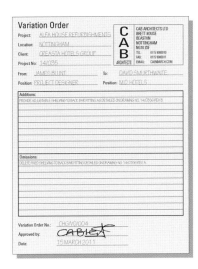

Figure 2.15 Variation order

Figure 2.16 Confirmation notice

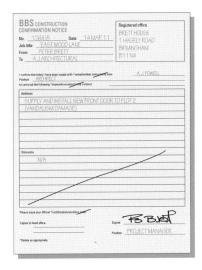

Figure 2.17 Daily report or site diary

Training and development records

Training and development is an important part of any job, as it ensures that employees have all the skills and knowledge that they need to do their work. Most medium to large employers will have training policies that set out how they intend to do this.

To make sure that they are on track and to keep records they will have a range of different documents. These will record all the training that an employee has undertaken.

Training can take place in a number of different ways:

* induction

* toolbox talks

* in-house training

* specialist training

* training or education leading to formal qualifications.

Scales used to produce construction drawings

When the plans for individual buildings or construction sites are drawn up they have to be scaled down so that they will fit on a manageable size of paper. It is important to remember that drawings are not sketches and that they are drawn to scale. This means that they are:

* exact and accurate

* in proportion to the real construction.

You can work out the dimensions by using the scale rule when measuring the drawings. There are several common scales used and the measurement is usually metric:

* 1:2500 – the drawing is 2,500 times smaller than the real object

* 1:100 – the drawing is 100 times smaller than the real object

* 1:50 – the drawing is 50 times smaller than the real object

* 1:20 – the drawing is 20 times smaller than the real object

* 1:10 – the drawing is 10 times smaller than the real object

* 1:5 – the drawing is 5 times smaller than the real object

* 1:2 – the drawing is 2 times smaller than the real object (also called 'half full size').

These drawings would clearly show the dimensions and these would be the actual measurements required (not the scaled-down measurements).

Selecting information

Various documents will provide you with the information you will need. The following examples show you how this works.

Location drawings and specifications

Location drawings are also known as block plans or site plans. The block plan shows the site presented as if you are looking down from above. It will show you where the site is in relation to other buildings and landmarks, such as roads.

The site plan will give you better detail of the site itself. It will contain measurements of the exact dimensions of the plot of land. It will also show you the routes of services and drainage.

The specifications are produced alongside the location drawings of the site. After giving you a brief description of the site, which includes the address, you will find:

* information about any services running into and through the site and whether or not they are connected or need to be connected

* whether there are restrictions about access or working hours on site

* materials that will be needed in order to carry out jobs on site (this will contain quite a lot of detail and will tell you what types of materials are needed, their size, quality and other technical details)

* information about the required workmanship, including what work needs to be done, what quality of work is expected and what the final finish should look like.

Schedules

Schedules can be quite big documents if you are working on a large site with lots of tasks to be done. There is usually a schedule for each different type of job. The schedules are there to record design information. They ensure that you do not accidentally use the wrong components or fittings.

Schedules cover all types of job, such as the types of doors, windows, joinery, heating components and other specific features.

Schedules usually have suitable drawings and usually a floor plan showing where the different features will appear.

PRACTICAL TIP

The same schedule can be used by several trades. For example, a window as far as a bricklayer is concerned means creating an opening that is large enough to take the window that has been selected. For the carpenter the window type and size is important, as they may have to construct a window of a specific size with particular openings and types of glass.

DETERMINE QUANTITIES OF MATERIALS

Working out the quantity and cost of resources that are needed to do a particular job is, perhaps, one of the most difficult tasks. In most cases you or the company you work for will be asked to provide a price for the work. It is generally accepted that there are three ways of doing this:

* **An estimate** is an approximate cost produced from the information available before construction begins.

* **A quotation** is a fixed price.

* **A tender** is a bid for the job at your price.

These three ways of costing are very different and each of them has its own problems.

Checking deliveries of building materials

It is important that all deliveries are thoroughly checked as they arrive. Your suppliers will need to have access to the site. It is important to inform suppliers whether the site has any access problems. Construction materials usually arrive on site on large and heavy lorries, so always check if the ground they will have to cross is soft or uneven and warn them if this is the case. Trees and overhead wires could also be a problem, as could finding space to reverse and turn the lorry.

If you are expecting a delivery to arrive, you should be prepared for it. This means ensuring there is:

* clear access to the site

* somewhere ready to store the materials and equipment being delivered

* enough help on hand to assist moving the materials from the delivery point to the storage area.

There are two documents that you will need in order to check the delivery:

* your order, which is a purchase order or confirmation of the materials and equipment that you have ordered from the supplier

* a delivery note, which should be handed to you by the delivery driver.

The first thing to do is check that the two documents match. If they do then what is on the delivery truck should be what you ordered. You now need to check and tick off each item on the delivery note. It needs to be:

* the right specification
* the right quantity

* the right size
* undamaged.

You should not accept items that do not match these four points. You should not sign the delivery ticket or delivery note unless you are satisfied with what has been delivered.

Methods used to estimate quantities

Numerous factors determine the cost of a construction project, whatever its size, but getting the best price on the ideal quantities should not be guesswork. Some possible considerations, according to the size and significance of the project, would be:

* the availability of labour and materials

* the lead time on materials

* the economy and borrowing rates

* the time of year

* cost of plant

* the duration of the project – really large projects can go on for years.

Obviously experience means that you can more quickly estimate the quantities of materials that will be needed on particular (small) construction projects. This is also true of working out the best place to buy materials and how much the labour costs will be to get the job finished.

Many businesses will use the *Hutchins UK Building Blackbook* (published by Franklin-Andrews), which provides a construction cost guide. It breaks down all types of work and shows an average cost for each of them.

Computerised estimating packages are available, which will give a comprehensive detailed estimate that looks very professional. This will also help to estimate quantities and timescales.

The alternative is of course to carry out a numerical calculation. So it is important to have the right resources upon which to base these calculations. These could be working drawings, schedules or other documents.

Usually this involves making additions, subtractions, multiplications and divisions. In order to work out the amount of materials you will need for a construction project you will need to know some basic information:

* What does the job entail? How complex is it, and how much labour is required?

* What materials will be used?

* What are the costs of the materials?

Measurement

The standard unit for measurement is the metre (m). There are 100 centimetres (cm) and 1,000 millimetres (mm) in a metre. It is important to remember that drawings and plans have different scales, so these need to be converted to work out quantities of materials.

The most basic thing to work out is length, from which you can calculate perimeter, area and then volume, capacity, mass and weight, as can be seen in Table 2.3.

Measurement	Explanation
Length	This is the distance from one end to the other. This could be measured in metres or milimetres, depending on the job.
Perimeter	This is the total distance around the outside of a shape. For example, you might need to know the length of the perimeter around a site to work out how much security fencing you need before work starts. You can work out the perimeter by adding the lengths of each side of the shape together. For most jobs, perimeter will be measured in metres (see Fig 2.20).
Area	This is the amount of surface a shape covers. For example you might need to work out the area of a room or a wall to calculate what quantity of materials you will need. You can work out the area of a room by measuring the length and the width of the room and multiplying the two figures together. You can work out the area of a wall by measuring the length and the height of the wall and multiplying the two figures together. For most jobs, area will be measured in square metres (m^2) (see Fig 2.21).
Volume and capacity	This shows how much space is taken up by an object or room. You can work out the volume of a room by multiplying the width by the length and then by the height. For most jobs, volume will be measured in cubic metres (m^3). Capacity works in exactly the same way as volume, but instead of showing the figure as cubic metres you may show it as litres (l). This is ideal if you are trying to work out the capacity of a water tank or a garden pond.
Mass or weight	Mass is measured usually in kilograms or in grams. Mass is the actual weight of a particular object, such as a brick.

Table 2.3

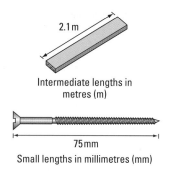

2.1 m

Intermediate lengths in metres (m)

75 mm

Small lengths in millimetres (mm)

Figure 2.18 Length in metres and millimetres

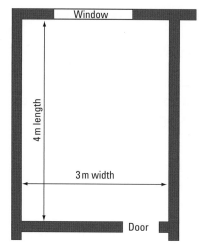

Figure 2.19 Measuring area and perimeter

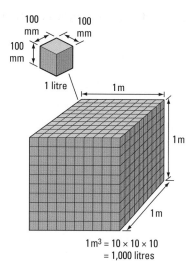

$1 m^3 = 10 \times 10 \times 10$
= 1,000 litres

Figure 2.20 Relationship between volume and capacity

Formulae

These can appear to be complicated, but using formulae is essential for working out quantities of materials. Each of the formulae is related to different shapes. In construction you will often have to work out quantities of materials needed for odd shaped areas.

Area

To work out the area of a triangular shape, you use the following formula:

$$\text{Area (A)} = \text{Base (B)} \times \frac{\text{Height (H)}}{2}$$

So if a triangle has a base of 4.5 and a height of 3.5 the calculation is:

$$4.5 \times \frac{3.5}{2}$$

$$\text{Or } 4.5 \times 3.5 = \frac{15.75}{2} = 7.875\,m^2$$

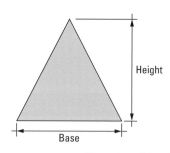

Figure 2.21 Triangle

Height

If you want to work out the height of a triangle you switch the formulae around. To give:

$$\text{Height} = 2 \times \frac{\text{Area}}{\text{Base}}$$

Perimeter

To work out the perimeter of a rectangle we use the formula:

$$\text{Perimeter} = 2 \times (\text{Length} + \text{Width})$$

It is important to remember this because you need to count the length and the width twice to ensure you have calculated the total distance around the object.

Circles

To work out the circumference or perimeter of a circle you use the formula:

$$\text{Circumference} = \pi \text{ (pi)} \times \text{diameter}$$

π (pi) is always the same for all circles and is 3.142.

Diameter is the length of the widest part and is twice the radius.

If we know the circumference and need to work out the diameter of the circle the formula is:

$$\text{Diameter} = \frac{\text{circumference}}{\pi \ (\text{pi})}$$

For example if a circle has a circumference of 15.39 m then to work out the diameter:

$$\frac{15.39}{3.142} = 4.89 \text{ m}$$

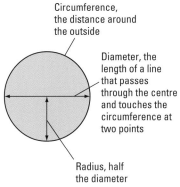

Figure 2.22 Parts of a circle

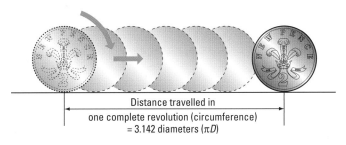

Distance travelled in
one complete revolution (circumference)
$= 3.142$ diameters (πD)

Figure 2.23 Relationship between circumference and diameter

Complex areas

Land, for example, is rarely square or rectangular. It is made up of odd shapes. Never be overwhelmed by complex areas, as all you need to do is break them down into regular shapes.

By accurately measuring the perimeter you can then break down the shape into a series of triangles or rectangles. All that is then necessary is to work out the area of each of the shapes within the overall shape and then add them together.

Shape	Area equals	Perimeter equals
Square	AA (or A multiplied by A)	4A (or A multiplied by 4)
Rectangle	LB (or L multiplied by B)	2(L+B) (or L plus B multiplied by 2)
Trapezium	$\frac{(A + B)H}{2}$ (or A plus B multiplied by H then divided by 2)	A+B+C+D

Shape	Area equals	Perimeter equals
Triangle	$\dfrac{BH}{2}$ (or B multiplied by H and then divided by 2)	A + B + C
Circle	πR^2 (or Pi (3.142) × R × R)	πD or 2πR (or Pi (3.142) × D or 2 × 3.142 × R)

Table 2.4

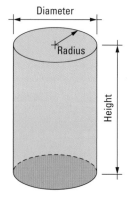

Figure 2.24 Cylinder

Volume

Sometimes it is necessary to work out the volume of an object, such as a cylinder or the amount of concrete needed. All that needs to be done is to work out the base area and then multiply that by the height.

For a concrete area, if a 1.2m square needs 3m of height then the calculation is:

$$1.2 \times 1.2 \times 3 = 4.32\,m^3$$

To work out the volume of a cylinder you need to know the base area × the height. The formula is:

$$\pi r^2 \times H$$

So if a cylinder has a radius (r) of 0.8 and a height of 3.5m then the calculation is:

$$3.142 \times 0.8 \times 0.8 \times 3.5 = 7.038\,m^3$$

Pythagoras

Pythagoras' theorem is used to work out the length of the sides of right angled triangles. It states that:

In all right angled triangles the square of the longest side is equal to the sum of the squares of the other two sides (that is, the length of a side multiplied by itself).

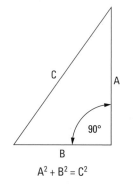

$$A^2 + B^2 = C^2$$

Figure 2.25 Pythagoras' theorem

Measuring materials

Using simple measurements and formulae can help you work out the amount of materials you will need. This is all summarised in the following table.

Material	Measurement
Timber	To work out the linear run of a cubic metre of timber of a given cross sectional area, divide a square metre by the cross sectional area of one piece.
Flooring	To work out the amount of flooring for a particular area metres^2 multiply the width of the floor by the length of the floor.
Stud walling, rafters and joists	Measure the distance that the stud partition will cover then divide that distance by a specified spacing and add 1. This will give you the number of spaces between each stud.
Fascias, barges and soffits	Measure the length and then add 10% for waste; however, this will depend on the nearest standard metric size of timber available.
Skirting, dado, picture rails and coving	You need to work out the perimeter of the room and then subtract any doorways or other openings. Again, add 10% for waste.
Bricks and mortar	Half brick walls use 60 bricks per metre squared and one brick walls use double that amount. You should add 5 per cent to take into account any cutting or damage. For mortar assume that you will need 1 kg for each brick.

Table 2.5

How to cost materials

Once you have found out the quantity of materials necessary you need to find out the price of those materials. It is then simply a case of multiplying those prices by the amount of materials actually needed.

Materials and purchasing systems

Many builders and companies will have preferred suppliers of materials. Many of them will already have negotiated discounts based on their likely spending with that supplier over the course of a year. The supplier will be geared up to supply them at an agreed price.

In other cases builders may shop around to find the best price for the materials that match the specification. It is not always the case that the lowest price is necessarily the best. All materials need to be of a sufficient quality. The other key consideration is whether the materials are immediately available for delivery.

DID YOU KNOW?

Many businesses that fail do so as a result of not working out their costs properly. They may have plenty of work but they are making very little money.

CASE STUDY

LAING O'ROURKE

What you learned at school comes in handy for work

Joshua Richardson is an apprentice in his third year at Laing O'Rourke in Leeds.

'I got an A in Resistant Materials (Design and Technology) at school and then decided it was a good idea to apply for a joinery apprenticeship. I applied for a few apprenticeships and eventually got one with Laing O'Rourke. They gave me a phone interview, and then I had to go to an assessment day where we did things like spatial tests, group activities and a 10-minute presentation. I'm glad I got to do similar tests with my other applications, because if this was the first place I'd done them, maybe I wouldn't have made it.

You had to have GCSEs in Maths, English and Science at A–C. There's some people at college who didn't have these so they have to do Functional Skills on top of everything.

It's important because you do use your Maths and English skills at work every day.

Not only do you need to use your English for college work-based evidence, but also just talking to people. I found that having done talks and presentations at school, it really helped communication skills, and I can talk quite easily with different people now.

With Maths, you definitely need that every day. When you're doing any job, you need to work out how to use a measuring tape, and what kind of Maths you're going to have to use. You have to measure up properly and know each calculation you'll need. It's all on your tape, but you've got to think about it and add it all together, and it's especially important to get it right when you're cutting.

If you're not too good with numbers, you just have to practise it – it's something that can come to you, and not everyone gets it straight away. Every so often, a guy will shout out a couple of measurements and say, "Add that!" So it's not as if you get to pull out your calculator every time.'

It is vital that suppliers are reliable and that they have sufficient materials in stock. Delays in deliveries can cause major setbacks on site. It is not always possible to warn suppliers that materials will be needed, but a well-run site should be able to anticipate needed materials and put the orders in within good time.

Large quantities may be delivered direct from the manufacturer straight to site. This is preferable when dealing with items where consistency is essential.

RELAYING INFORMATION IN THE CONSTRUCTION ENVIRONMENT

Communication is all about passing on accurate information. No matter who you are communicating with, you must understand what is being asked. You also need to be able to give them a clear answer. Sometimes you do not have the information or it is not your decision to make. In these cases you will need to take a message.

Whatever the situation, you always need to be positive and efficient. You also need to be clear. Poor communication and negative communication nearly always lead to confusion, delays and extra costs.

Basic content and requirements for recording a message

Message taking is an important skill. You are the way in which information is passed from one person to another. Providing you remember to pass the message on it may not need to be written down. In many cases, however, messages can be complicated and these do need to be written down. In fact it is a wise plan to record the fact that the message was received in the first place:

* Note the date and time that the message was received.

* Clearly write down the actual message. Someone will have to read this, so make sure it is legible.

* Write down the person's name and their contact details.

If you are taking a message and you are in the site office then there may be a telephone answering pad to hand, or perhaps sticky notes.

If the questions are complicated it will be best to get them to call back or to promise them that the person with the knowledge will call them back.

> **PRACTICAL TIP**
>
> Many people who leave a message will tell you that it is urgent. It is not for you to decide whether it is or not. It is just down to you to pass the message on and mark it urgent if they have said so.

Positive and negative communication

Construction is an industry that relies on communication.

Positive communication means:

* being courteous and respectful when you are talking to others

* being considerate, particularly if the other person is under pressure

* listening to what others are saying

* being clear on key points

* keeping a sense of humour.

Showing these positive communication skills should mean that others will show positive communication towards you.

On the other hand, if you are:

* rude and disrespectful

* unwilling to listen or pay attention

* incapable of making a decision

* bad tempered

then you are communicating in a negative way. This could lead to confusion, arguments and problems.

Clear and effective communication

Communication in all types of work is essential. It needs to be clear and to the point, as well as accurate. Above all it needs to be a two-way process. This means that any communication you have with anyone must be understood. Think before communicating and never assume that someone understands you unless they have confirmed that they do. Negative or poor communication can damage the confidence that others have in you to do your job.

In construction work everything is about time and following strict instructions and specifications. Failing to communicate will always cause confusion, extra cost and delays. In an industry such as this these are unacceptable and very easy to avoid.

Good communication means efficiency and achievement.

COMMUNICATING WITH OTHERS IN THE CONSTRUCTION ENVIRONMENT

Communication can be split into two different types:

* **Verbal communication** includes face-to-face conversations, discussions in meetings or performance reviews and talking on the telephone.

* **Written communication** includes all forms of documents, from letters and emails to drawings and work schedules.

Each of these forms of communication needs to be clear, accurate and designed in such a way as to make sure that whoever has to use it or refer to it understands it.

Communicating in the appropriate way with others

Each construction job will require the services of a team of professionals. They will need to be able to work and communicate with one another. Each has different roles and responsibilities. They can be broken down into three particular groups:

* on site
* off site
* visitors.

These are described in Tables 2.6, 2.7 and 2.8.

On site

Role	Responsibilities
Apprentices	They can work for any of the main building services trades under supervision. They only carry out work that has been specifically assigned to them by a trainer, a skilled operative or a supervisor.
Skilled or trade operative	A specialist in a particular trade, such as bricklaying or carpentry. They will be qualified in that trade, or working towards their qualification
Unskilled operatives	Also known as labourers, these are entry level operatives without any formal training. They may be experienced on sites and will take instructions from the supervisor or site manager.
Building services engineers	They are involved in the design, installation and maintenance of heating, water, electrics, lighting, gas and communications. They work either for the main contractor or the architect and give instruction to building services operatives.
Building services operatives	They include all the main trades involved in installation, maintenance and servicing. They take instruction from the building services engineers and work with other individuals, such as the supervisor and charge-hand.
Sub-contractor	They carry out work on behalf of the main contractor and are usually specialist tradespeople or professionals, such as electricians. Essentially, they provide a service and are contracted to complete their part of the project.
Charge-hand	This person supervises a specific trade, such as carpenters and bricklayers.
Site manager	This person runs the construction site, makes plans to avoid problems and meet deadlines, and ensures all processes are carried out safely. They communicate directly with the client.
Supervisor	The supervisor works directly for the site manager on larger projects and carries out some of the site manager's duties on their behalf.
Health and safety officer	This person is responsible for managing the safety and welfare of the construction site. They will carry out inspections, provide training and correct hazards.

Table 2.6

Off site

Role	Responsibilities
Client	The client, such as a local authority, commissions the job. They define the scope of the work and agree on the timescale and schedule of payments.
Customer	For domestic dwellings, the customer may be the same as the client, but for larger projects a customer may be the end user of the building, such as a tenant renting local authority housing or a business renting an office. These individuals are most affected by any work on site. They should be considered and informed, with a view to them suffering as little disruption as possible.
Architect	They are involved in designing new buildings, extensions and alterations. They work closely with clients and customers to ensure the designs match their needs. They also work closely with other construction professionals, such as surveyors and engineers.

Consultant	Consultants such as civil engineers work with clients to plan, manage, design or supervise construction projects. There are many different types of consultant, all with particular specialisms.
Main contractor	This is the main business or organisation employed to head up the construction work. They organise the on-site building team and pull together all necessary expertise. They manage the whole project, taking full responsibility for its progress and costs.
Clerk of works	This person is employed by the architect on behalf of a client. They oversee the construction work and ensure that it represents the interests of the client and follows agreed specifications and designs.
Quantity surveyor	Quantity surveyors are concerned with building costs. They balance maintaining standards and quality against minimising the costs of any project. They need to make choices in line with Building Regulations. They may work either for the client or for the contractor, and clients and contractors may both have quantity surveyors on site.
Estimator	Estimators calculate detailed cost breakdowns of work based on specifications provided by the architect and main contractor. They work out the quantity and costs of all building materials, plant required and labour costs.
Supplier/wholesaler contracts manager	They work for materials suppliers or stockists, providing materials that match required specifications. They agree prices and delivery dates.

Table 2.7

Visitors

Site visitor	Role and responsibility
Training officers and assessors	These people work for approved training providers. They visit the site to observe and talk to apprentices and their mentors or supervisors. They assess apprentices' competence and help them to put together the paperwork needed to show evidence of their skills.
Building control inspector	This person works for the local authority to ensure that the construction work conforms to regulations, particularly the Building Regulations. They check plans, carry out inspections, issue completion certificates, work with architects and engineers and provide technical knowledge on site.
Water inspector	This person carries out checks of plumbing and drainage systems on construction sites.
Health and Safety Executive (HSE) inspector	An HSE inspector from the local authority can enter any workplace without giving notice. They will look at the workplace, the activities and the management of health and safety to ensure that the site complies with health and safety laws. They can take action if they find there is a risk to health and safety on site.
Electrical services inspector	Inspectors are approved by the National Inspection Council for Electrical Installation Contracting. They check all electrical installation has been carried out in accordance with legislation, particularly Part P of the Building Regulations.

Table 2.8

Maintaining good working relationships

It is important to have a good working relationship with colleagues at work. An important part of this is to communicate in a clear way with them. This helps everyone understand what is going on and what decisions have been made. It also means being clear. Most communication with colleagues will be verbal (spoken). Good communication means:

* cutting out mistakes and stoppages (saving money)

* avoiding delays

* making sure that the job is done right the first time and every time.

Equality and diversity in communication

Equality and diversity is not simply about treating everyone in the same way. It is actually recognising that people are different. Each of us is unique. This could mean that we might have a different culture, be of different ages or follow different religions. It might refer to our marital status or gender, our sexual orientation or our first language.

In all your actions and your communications you should:

* recognise and respect other people's backgrounds

* recognise that everyone has rights and responsibilities

* not harass or be offensive and use language or behaviour that discriminates.

You should also remember that not everyone's first language will be English so they may not understand everything or be able to communicate clearly with you. You might also find that some colleagues may have hearing impairments (or may not hear what you're saying because they are in a noisy environment). It's best to use simple language and check that both you and the person you're communicating with have understood what you need to know.

Figure 2.26 A water inspection

CASE STUDY

Using writing and maths in the real world

Gary Kirsop, Head of Property Services says:

'People seem to think that trades are all about your hands, but it's more than that. You're measuring complicated things – all the trades need to have about the same technical level for planning, calculation and writing reports. You need that level to get through your exams for the future too. When you have one day a week in college, but four days a week working with customers in the real world, without communications skills, it would all fall apart. You have to understand that people come from different backgrounds and that they have their own communication modes. Having good GCSEs will really help you get by in the trade.'

TEST YOURSELF

1. If a drawing is at a scale of 1:500, each millimetre in the drawing represents how much on the ground?

   a. 1 m

   b. 1 mm

   c. 500 mm

   d. 500 m

2. What is the other term used to describe an orthographic projection?

   a. First angle

   b. Second angle

   c. Third angle

   d. Isometric

3. What is a variation order?

   a. A list of all materials that have been delivered to site

   b. A document showing work that has been carried out without a prior estimate

   c. A document that confirms any verbal instructions

   d. A document provided by the architect to the builder to show any changes to the original job

4. On a drawing, if you were to see the letters FDN, what would that mean?

   a. The signature of the architect

   b. Foundation Design Network

   c. Foundations

   d. Full distance

5. If a drawing is at a scale of 1:5, how many times smaller is the drawing than the real object?

   a. 5 times

   b. 50 times

   c. Half the size

   d. 500 times

6. Which of the following values is pi?

   a. 3.121

   b. 3.424

   c. 3.142

   d. 3.421

7. Which document is used to give detailed sets of requirements that cover the construction, features, materials and finishes?

   a. Work programme

   b. Purchase order

   c. Policy document

   d. Job specification

8. How do you work out the amount of flooring necessary for a room?

   a. Divide width by length

   b. Add width to length

   c. Multiply length by height

   d. Multiply width by length

9. Which individual on a typical site would sign off timesheets?

   a. Architect

   b. Site manager/supervisor

   c. Delivery person

   d. Customer

10. Which are the two main types of communication?

    a. Verbal and written

    b. Telephones and emails

    c. Meetings and memorandum

    d. Plans and faxes

Chapter 3

Unit CSA–L1Core03
KNOWLEDGE OF CONSTRUCTION TECHNOLOGY

LEARNING OUTCOMES

LO1: Know about foundation construction

LO2: Know about floor construction

LO3: Know about wall construction

LO4: Know about roof construction

LO5: Know about utilities and services within construction

LO6: Know about sustainability within construction

INTRODUCTION

The aim of this chapter is to:

* help you understand the range of building materials used within the construction industry

* help you understand their suitability to the construction of modern buildings.

* help you understand the role of sustainability in the construction industry

* help you to be aware of different construction methods.

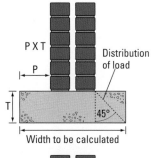

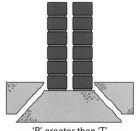

'P' greater than 'T' leads to shear failure

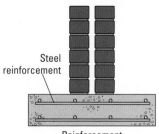

Reinforcement used to reduce 'T'

Figure 3.1 Foundation properties

FOUNDATION CONSTRUCTION

Foundations are a primary element of a building, and form part of the substructure (the element of the building which is below ground and cannot be seen once the structure has been completed). Foundations spread the load of the superstructure (the visible part of the completed building), transferring it to the subsoil ground below. They provide structural stability and help to prevent damage to the building in the event of ground movement. They can range from a concrete strip to pre-cast reinforced concrete-driven piles.

Purpose of foundations

Foundations are designed to counteract factors such as ground movement, which could damage the building. It is important to work out the necessary width of foundation. This depends on the total load of the structure and the load-bearing capacity of the ground or subsoil on which the building is being constructed:

* Wide foundations are used when the construction is on weak ground, or the superstructure will be heavy.

* Narrow foundations are used when the subsoil is capable of carrying a heavy weight, or the building is a relatively light load.

The load that is placed on strip and pad foundations spreads into the ground at 45°. Shear failure will take place if the thickness of the foundations is less than the projection of the wall or column face on the edge of the foundations. This is what leads to subsidence (the ground under the structure sinking or collapsing).

As we will see in this section, the depth of the foundation is dependent on the load-bearing capacity of the subsoil. As a general rule of thumb, foundations should be 200 mm to 300 mm thick.

Different types of foundation

Generally there are several different types of foundation, which can be seen in Fig 3.2.

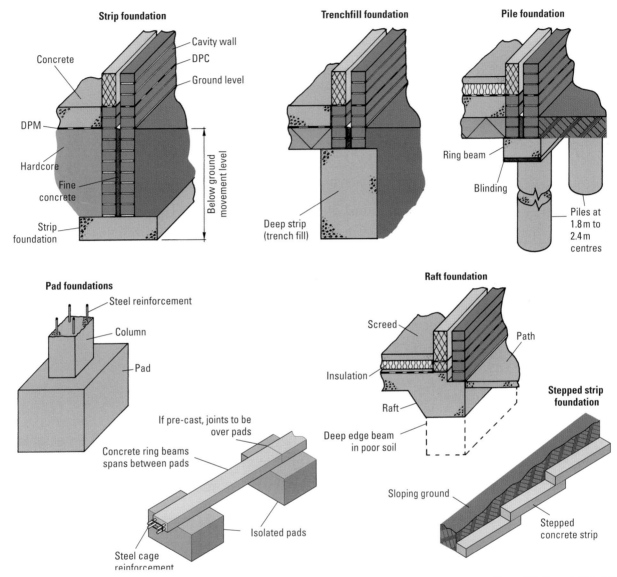

Figure 3.2 Foundation types

* The traditional **strip foundation** is quite narrow and tends to be used for low-rise buildings and dwellings. A thin strip of concrete is laid and then brick or block is built up to the DPC level. These can be reinforced where the ground is weak. They can also be stepped on sloping ground, in order to cut down on the amount of excavation needed. It can also be deep, which uses more concrete but reduces the number of bricks used below ground level. An alternative to deep strip foundations is a trench fill foundation.

* Trench fill foundations are constructed by digging a narrow trench to the foundation depth and then filled with concrete. This reduces the labour and materials required to lay the foundation, as no bricks or blocks have to be laid into the trench. Trench fill:

* reduces the need to have a wide foundation

* reduces construction time

* speeds up the construction of the foundation.

* **Pad foundations** tend to be used for structures that have either a concrete or a steel frame. The pads are placed to support the columns, which transfer the load of the building into the subsoil.

* **Pile foundations** tend to be used for high-rise buildings or where the subsoil is unstable. Holes are bored into the ground and filled with concrete or pre-cast concrete, steel or timber posts are driven into the ground. These piles are then spanned with concrete ring beams with steel reinforcement so that the load of the building is transferred deeper into the ground below. Pile foundations can be short or long depending on how high the building is or how bad the soil conditions are.

* **Raft foundations** are used when there is a danger that the subsoil is unstable. A large concrete slab reinforced with steel bars is used to outline the whole footprint of the building. It has an edge beam to take the load from the walls which is transferred over the whole raft. This means that the building effectively 'floats' on the ground surface on top of the concrete raft.

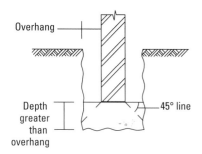

Figure 3.3 Unreinforced strip foundation

Selecting a foundation

One of the first things that a structural engineer will look at when they investigate a site is the nature of the soil, the ground conditions, the likelihood of ground movement and issues such as the water table (where groundwater begins).

Table 3.1 shows different types of subsoil and how they can affect the choice of foundation.

Subsoil type	Characteristics
Rock	High load bearing but there may be cracks or faults in the rock, which could collapse.
Granular	Medium to high load bearing and can be compacted sand or gravel. If there is a danger of flooding the sand can be washed away.
Cohesive	Low to medium load bearing, such as clay and silt. These are relatively stable, but there may be problems with water.
Organic	Low load bearing, such as peat and topsoil. There is also a great deal of air and water present in the soil so organic material must be removed before starting the foundations.

Table 3.1

The ground may move, particularly if the conditions are wet, extremely dry or there are extremes of temperature. Clay, for example, will shrink in the hot summer months and swell up again in the wet winter months. Frost can affect the water in the ground, causing it to expand (frost heave).

Ground movement is also affected by the proximity of trees and large shrubs. They will absorb water from the soil, which can dry out the subsoil. This causes the soil underneath the foundations to collapse. This may not be a problem until the tree or shrub is removed or new ones planted.

As we have seen above, the other key factor when selecting a foundation is the end use of the building:

* Strip foundations – this is the most common and cheapest type of foundation. It is a strip of concrete that runs under the load-bearing walls. The actual depth and width of the strip depends on the ground and the load from the building. They are used for low to medium rise domestic and industrial buildings, as the load bearing requirement will not be high.

* Piled foundations – tend to be used for high rise buildings or where subsoil is unstable if the ground directly under the building is weak or unstable then concrete or steel piles can be driven through this weak ground and into more solid ground beneath it. Reinforced concrete ring beams are placed over the piles as a direct support for the building.

* Raft foundations – these are very expensive and are only ever really used when the ground on which the building is being constructed is very soft. It is also sometimes used when the ground across the area is likely to react in different ways because of the weight of the building. In areas of the UK where there has been mining, for example, raft foundations are quite common, as the building could subside.

Concrete

Concrete is the most common material used for foundations as it is strong and durable. It is usually cast directly on site.

The concrete needs to be poured into the foundation with some care. The size of the foundation will usually determine whether the concrete is actually mixed on site or brought in, in a ready-mixed state, from a supplier. For smaller foundations a concrete mixer and wheelbarrows are usually sufficient. The concrete is then poured into the foundation using a chute (a long trough with a rounded bottom and open ends that directs the concrete to where it is needed).

Concrete consists of both fine and coarse aggregate, along with water.

Aggregates

Aggregates are basically fillers. Coarse aggregate is usually either crushed rock or gravel. The grains are 5mm or larger.

The fine aggregate fills up any gaps between the particles in the coarse aggregate.

Fine aggregate is usually sand that has grains smaller than 5mm.

Cement

Cement is an adhesive or binder. It is Portland stone, crushed, burnt and crushed again and mixed with limestone. The materials are powdered and then mixed together to create a fine powder, which is then fired in a kiln. It may be mixed with other materials for different purposes, such as creating masonry mortar.

Water

Potable water, which is water that is suitable for drinking, should be used when making concrete. The reason for this is that drinkable water has not been contaminated and it does not have organic material in it that could rot and cause the concrete to crack. The water mixes with the cement and then coats the aggregate. This effectively bonds everything together.

Additives

Additives, or admixtures, make it possible to control the setting time and other aspects of fresh concrete allowing you to have greater control over the concrete. They can:

- give you higher strength concrete
- provide protection against degradation of concrete or corrosion of reinforced concrete, which will weaken the structure
- speed up the time the concrete needs to set
- reduce the time the concrete takes to set
- provide protection against cracking as the concrete sets (by preventing shrinkage)
- improve the flow (workability) of the concrete
- improve the finish of the concrete
- provide hot or cold weather protection (a drop or rise in temperature can change the amount of time that concrete needs to set, so these admixtures compensate for that).

You might need this flexibility if, for example, the schedule or weather changes or the job has an unusual specification.

> **DID YOU KNOW?**
>
> Ordinary Portland cement is Portland stone crushed and burnt until all the water disappears. This is taken to site where the water is added again to reconstitute the stone to the form required.

Figure 3.4 Reinforcement using steel bars (or mesh)

Reinforcement

Steel bars or mesh can be used to give the foundation additional strength and support. It can also help to stop the foundation from cracking. Concrete is a good material under direct weight loads, but where concrete foundations are wide and parts of them are under additional tension there is a danger they may crack.

Natural and artificial stone

Both of these products can be placed over the top of plain foundations hardcore fill that is put into the substructure to fill the gap up to the ground floor level hardcore fill that is put into the substructure to fill the gap up to the ground floor level. The synthetic stone weighs far less than natural stone. The other advantage with the synthetic products is that the foundations do not need to be as substantial.

FLOOR CONSTRUCTION

A floor is a level surface that provides some insulation and carries any loads on it (for example, from furniture) and then to transfer those loads.

Ground floors also have additional purposes. They need to stop moisture from entering the building from the ground. They also need to prevent plant or tree roots from entering the building.

Ground floors

For ground floors there are two options:

* Solid – in contact with the ground

* Suspended – does not touch the ground and spans between walls in the building (effectively there is a void beneath the floor)

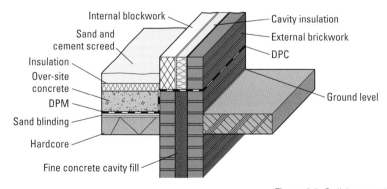

Figure 3.5 Solid ground floors

Floating floor

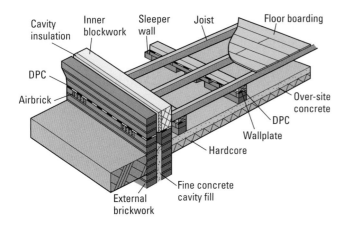

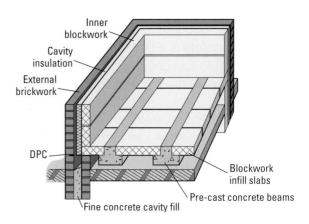

Figure 3.6 Suspended ground floors

The options for ground floors can be complicated because they need to perform several functions. While new builds don't tend to have timber joists and floorboards, extensions to existing buildings usually need to match existing construction styles. Suspended ground floors and traditional timber floors tend to be seen in older buildings. It is far more common to have solid ground floors, or to have timber floors over concrete floors, which are known as floating ground floors.

The key options are outlined in Table 3.2.

Type of floor	Construction and characteristics
Solid	The ground is compacted, and compacted hard core is used as the base, with a binding layer of sand, which is covered with a damp-proof membrane (dpm). A layer of insulation board, usually 100 mm thick, is then placed onto the dpm and concrete is poured on top. To provide a smooth finish for floor finishes a cement and sand screed is applied, usually after the building has been made watertight.
Suspended	Timber – a similar process to a solid ground floor is carried out but then, on top of this, dwarf walls or sleepers are built. These are used to support the timber floor. Air bricks are also added to provide necessary ventilation. Joists are then spaced out along the dwarf walls. A damp-proof course is inserted under the floor joists and then floorboards or sheets placed on top of the joists. Beam and block – concrete beams and lightweight concrete slabs or blocks are used to create the basic flooring. The beams are evenly spaced across the foundation and gaps between the beams are filled with blocks to form the floor. The blocks and beams are then insulated and it is finished off with a cement screed.
Floating	This is a timber construction which goes over the top of a solid concrete floor. Bearers are put down and then the boarding or sheets are fixed to the bearers. The weight of the boards themselves hold them in place.

Table 3.2

Upper floors

Usually for dwellings timber is used for these suspended floors. In industrial buildings beam and block or concrete floor slabs tend to be used.

Timber suspended upper floor

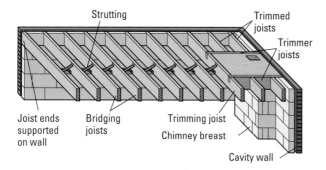

Concrete suspended upper floor

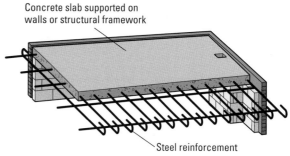

Figure 3.7 Upper floors

For dwellings, bridging joists (horizontal timbers that support the ceiling) are the most common joists used in suspended timber floors. These joists are supported at their ends by load-bearing walls. On the top of the joists, boarding or sheets provide the flooring for the room. Underneath the joists, plasterboard creates the basis of the ceiling for the room below.

When joists have to go into cavity walls (two walls with a hollow space between them) joist hangers are used (U-shaped metal brackets that are used to support the ends of floor joists). There are also complications when joists are in and around stairs and chimney breasts. Bridging joists are used so that these openings are not blocked. Openings in floors require the use of different types of joist called trimmers, trimming and trimmed joists. A trimmed joist is a shortened bridging joist. Any opening in a floor is treated in this way. When the span of a bridging joist exceeds 2 m then struts will be required in line with Building Regulations.

The voids between the floorboards and the plasterboard must be filled with insulation. This not only reduces heat loss, but can also reduce noise.

Concrete suspended floors are usually either cast on site or available as ready-cast units. They are effectively locked into the structure of the building by steel reinforcement. If the concrete floors are being cast on site then formwork is needed. Concrete floors are common because they offer greater load bearing capacity, have greater fire resistance and are more sound resistant.

KEY TERMS

Formwork

– this can also be known as shuttering. It is a temporary structure that supports and shapes wet concrete until it cures and is able to be self-supporting.

REED TIP
•••

Try to remember that an effective team can produce more combined that a person can do on their own. Working with other people is also good for your personal well-being.

WALL CONSTRUCTION

Walls have a number of different purposes as they:

* hold up the roof

* provide protection against the elements

* keep the occupants of the building warm

* divide the building into rooms, providing privacy and different spaces.

External walls

Many buildings now have cavity walls which means:

* The outside wall is a wet one because it is exposed to the elements outside the building.

* The internal wall is dry but it needs to be kept separate from the outside wall by a cavity.

* The cavity or gap acts as a barrier against damp and also provides some heat insulation.

* The cavity can be completely filled or part-filled depending on the insulation value required by Building Regulations.

Internal walls

Internal walls divide up the space within the building. These do not have all of the demands of the external walls. They are less likely to be load bearing and they do not have to be insulated so are, therefore, thinner. (However, they are commonly insulated in areas such as the toilet or party walls in semi-detached or terrace construction.) They can be brick or block (particularly if they are load bearing), which is then covered with plaster. Alternatively they can be a timber or metal framework, known as stud work, which is covered plasterboard, to form a wall.

Different types of wall construction and structural considerations

In addition to walls being external or internal, they can also be classed as being load bearing or non-load bearing.

Internal walls can be either load bearing or non-load bearing. In both external and internal load bearing walls, any gaps or openings for windows or doors have to be bridged. This is achieved by using either arches or lintels. These support the weight of the wall above the opening.

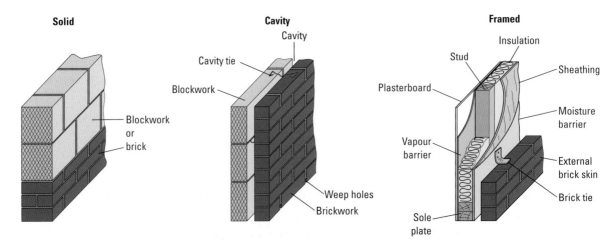

Figure 3.8 Some examples of external wall construction

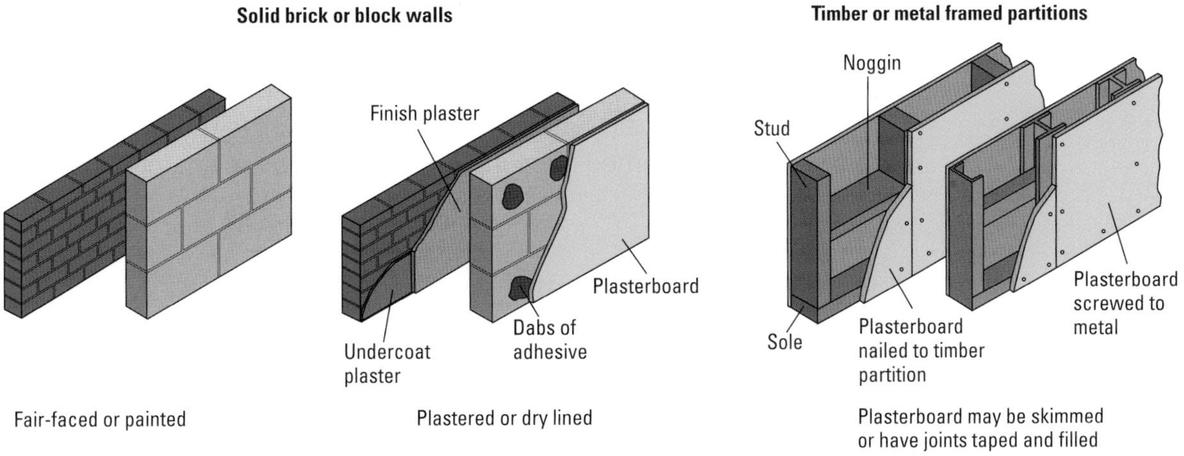

Solid brick or block walls

Timber or metal framed partitions

Finish plaster

Plasterboard

Undercoat plaster

Dabs of adhesive

Fair-faced or painted

Plastered or dry lined

Noggin

Stud

Sole

Plasterboard nailed to timber partition

Plasterboard screwed to metal

Plasterboard may be skimmed or have joints taped and filled

Figure 3.9 Some examples of internal wall construction

Solid walls

In modern builds construction of solid external walls is quite rare. External solid walls tend to be much thinner and made from lightweight blocks in modern builds. They will have some kind of waterproof surface over the top of them, which could be made of render, or plastic, metal or timber cladding.

Cavity walls

As we have seen, cavity walls have an outer and an inner wall and a cavity between them. Usually solid walling, or blockwork, is built up to ground level and then the cavity walling continues to the full height of the building. Alternatively a filled cavity wall is constructed up to ground level. Cavity walls are ideal for most buildings up to medium height.

Many industrial buildings have cavity walls for the lower part of the building and then have insulated steel panels for the top part of the building.

The usual technique is to have brick for the outer wall and an insulating block for the inner wall. The gap or cavity can then be partially filled with an insulation material.

Timber framed walls

Panels made of timber, or in some cases steel, are used to construct walls. They can either be load bearing or non-load bearing and can also be used for the outside of the building or for internal walling. Timber frames are also often clad in brickwork. The panels are solid structures and the spaces between the vertical struts (studs) and the horizontal struts (head or sole plates) are filled with insulation material.

Internal walls

Internal walls are either solid or framed. Solid walls can be made up from blocks or bricks. In many industrial buildings the blocks are actually exposed and can be left in their natural state or painted. In domestic buildings plasterboard is usually bonded to the surface and then plastered over to provide a smoother finish.

It is more common for domestic buildings to have timber or metal-framed internal walls, made from either timber or metal, which are known as stud partitions. These are exactly the same as other framed walling, but will usually have plasterboard fixed to them. They would then receive a skimmed coat of plaster to provide the smooth finish.

Damp-proof membrane (DPM) and damp-proof course (DPC)

Damp-proof membranes are installed under the concrete in ground floors in order to ensure that ground moisture does not enter the building. Effectively it waterproofs the building.

Damp-proof courses are a continuation of the damp-proof membrane. They are built into a horizontal course of either block or brickwork, which is a minimum of 150mm above the exterior ground level. DPCs are also designed to stop moisture from coming up from the ground, entering the wall and then getting into the building. The most common DPC is a polythene sheet damp-proof membrane, which comes in rolls the width of the blockwork or brickwork. In older buildings lead, bitumen or slate would have been used as a DPC.

ROOF CONSTRUCTION

In a country such as the UK, with a great deal of rain and snow, it makes sense for roofs to be pitched. Pitched means built at an angle. The idea is that the rain and snow falls down the angle and off the edge of the roof or into gutters rather than lying on the roof.

This is not to say that all roofs are pitched. In fact many domestic dwelling extensions have flat roofs. A great number of industrial buildings have entirely flat roofs. The problem with a flat roof is that it needs to be able to support itself, but as importantly it needs to be able to carry the additional weight of snow or rain. This means that large flat roofs may have to have steel sections (known as trusses) or even reinforced concrete and beams to increase their load-bearing capacity.

Roofs also provide stability to the walls by tying them together. As we will see, there are several different types of roof. These are usually identified by their pitch or shape.

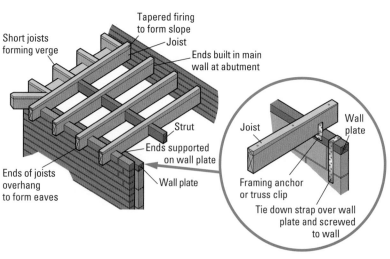

Short joists forming verge

Tapered firing to form slope

Joist

Ends built in main wall at abutment

Strut

Ends supported on wall plate

Ends of joists overhang to form eaves

Wall plate

Joist

Wall plate

Framing anchor or truss clip

Tie down strap over wall plate and screwed to wall

Figure 3.10 Flat roof structure

Types of roof construction

The roof is made up of the rafters and beams, Everything above the framework is regarded as a roof covering, such as slates, tiles and felt. Generally speaking, roofs are either pitched or flat. This will depend on the angle or slope of the roof.

Table 3.3 outlines some of the key characteristics of different types of roof.

Flat	This is a roof that has a slope of less than 10°. Generally flat roofs are used for smaller extensions to dwellings and on garages. Traditionally they would have had bitumen felt, although it is becoming more common for fibreglass to be used.	 Figure 3.11
Mono-pitch	This is a roof that has a single sloping surface but is not fixed to another building or wall. The front and back walls could be different heights, or the other exposed surface of the roof is perpendicular.	 Figure 3.12
Double pitched	This is a roof that has two differently angled slopes. Usually the upper part of the roof has a fairly shallow pitch or slope and the lower part of the roof has a steeper slope.	 Figure 3.13
Couple roof	This is often called gable end and is one of the most common types of roof for dwellings. A gable is a wall with a triangular upper part. This supports the roof in construction using purlins. This means that the roof has two sloping surfaces, which come down from the ridge to the eaves.	 Figure 3.14
Hipped roof	Hipped roofs have slopes on three or four four sides. There are also hipped roofs with single, straight gables.	 Figure 3.15
Lean-to	A lean to is similar to a mono-pitched roof except it is abutted to a wall. The slope is greater than 10°. The higher part of the roof is fixed to a higher wall.	 Figure 3.16

Table 3.3 Different types of roof

Roofing components

Each part of a roof has a specific name and purpose. Table 3.4 explains each of these individual features.

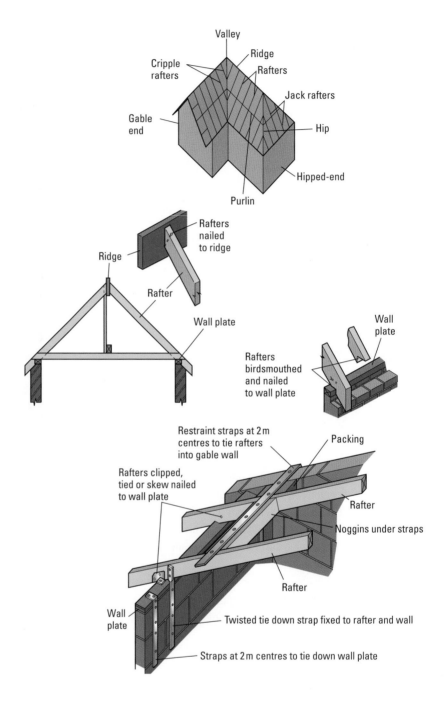

Figure 3.17 Traditional cut roof details

Roof feature	Description
Ridge	This is the top of the roof and the junction of the sloping sides. It is the apex, where the rafters meet.
Purlin	This is a beam that supports the mid-span section of rafters.
Firings	These are angled pieces of timber that are placed on the rafters to create a slope.
Batten	Roof battens are thin strips, usually of wood, which provide a fixing point for either roofing sheets or roof tiles.
Tile	These are artificial products that can be made from clay, concrete or plastic. They are placed in regular, overlapping rows and fixed to the battens.
Fascia	This is a horizontal board that closes and protects the rafter ends and provides a fixing for guttering. It is also decorative as it covers the rafter ends. It is fixed to the ends of the rafters at eaves level and is both a decorative feature and a fixing for rainwater goods.
Wall plate	This is a piece of horizontal timber that is placed at the top of a wall at eaves level. It provides a fixing for joists or rafters.
Bracings	Roof rafters may need to be braced to make them more rigid and stable. These bracings prevent the roof from buckling.
Felt	Roofing felt has two elements – it has a waterproofing agent (bitumen) and what is known as a carrier. The carrier can be either a polyester sheet or a glass fibre sheet. Roofing felt tends to be used for flat roofs and for roofs with a shallow pitch.
Slate	Slate roofing tiles are natural products and are usually fixed to timber battens with double nails. They have a lifespan of between 80 and 100 years.
Flashings	Wherever there is a joint or angle on a roof, a thin sheet of either lead or another waterproof material is added. In the past this tended always to be made from lead. Many different types of flashing can now be used but all have the role of preventing water penetrating into joints. Flashings are normally found where roofs abut a wall or where chimneys protrude through a roof.
Rafter	Roof rafters are the main structural components of the roof. They are the framework. They rest on supporting walls. The rafters are set at an angle on sloped roofs or horizontal on a flat roof.
Apex	The apex is the highest point of the roof, usually the ridge line.
Soffit	Soffits are the lower part, or overhanging part, of the eaves. In other words they are the underside of the eaves.
Bargeboard	This is a functional and ornamental feature, which is fixed to the gable end of a roof in order to hide the ends of roof timbers and to support the verge details.
Eaves	These are the area found at the foot of the rafter. They are not always visible as they can be flush. In modern construction, the eaves have two parts: the visible eaves projection and the hidden eaves projection.

Table 3.4

Roof coverings

There are many different types of materials that can be used to cover the roof. Even tiles and slates come in a wide variety of shapes and sizes, along with colours and different finishes.

In many cases the type of roof covering is determined by the traditional and local styles in the area. Local authorities usually want roof coverings that are not too far from the common style in the area. This does not stop manufacturers from coming up with new ideas, however, which can add benefits during construction and during the use of the building (such as better insulation properties).

Table 3.5 outlines some of the more common types of roof covering and describes their main characteristics and use.

Roof covering	Description
Felt  Figure 3.18	Felt is used as a waterproof barrier. Internal felt is rolled over the top of the rafters. The strips are overlapped to provide a permanent waterproof barrier. They are then battened down and another roof covering, such as slate or tile, placed over the top of them. For flat roofs, felt is used as the external roof covering and is covered in a waterproof material, such as bitumen.
Slate 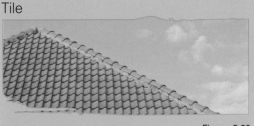 Figure 3.19	Slate is a flat, natural substance, which is laid onto the battens with each slate tile overlapping the top of the slate in the row directly below it. The slate tiles are either nailed or hooked into place.
Tile Figure 3.20	There is a huge variety of roofing tiles, made from clay, ceramics or concrete. They are designed and moulded so that they overlap with one another and are fixed to the roof in a similar way to slate tiles.
Metals  Figure 3.21	There are many different types of metal roof covering, such as corrugated sheets, flat sheets, box profile sheets or even sheets that have a tile effect. The metal is galvanised and plastic coated to provide a durable and long-lasting waterproof surface.

Table 3.5

CASE STUDY

South Tyneside Homes

South Tyneside Council's
Housing Company

How to impress in interviews

Andrea Dickson and Gillian Jenkins sit on the interview panels for apprenticeship applications at South Tyneside Homes.

'Interviews are all about the three Ps: Preparation, Presentation and Personality.

An applicant should turn up with some knowledge about the apprenticeship programme and the company itself. For example, knowing how long it is, that they have to go to college and to work – don't say, "I was hoping you'd tell me about it"! If they've done a bit of research, it will show through and work in their favour – especially if they can explain why it is that they want to work here.

It sets them up for the interview if they come in smartly dressed. We're not marking on that, but it does show respect for the situation. It's still a formal process and, although we try to make them feel at ease as much as we possibly can, there's no getting away from the fact that they're applying for a job and it is a formal setting.

The interviews are a chance to tell the company about themselves: what they do in their spare time, what their greatest achievements have been and why. Applicants should talk about what interests them; for example, are they really interested in becoming a joiner or is that something their parents want them to do? An apprenticeship has to be something they want to do – if they have enthusiasm for the programme, then they'll fly through it. If not, it's a very long three to four years. Without that passion for it, the whole process will be a struggle; they'll come in late to work and even fail exams.

We also talk to them about any customer service experiences they've had, working in a team, project working (for example, a time you had to complete a task and what steps you took), as well as asking some questions about health and safety awareness.'

SUPPLY OF UTILITIES AND SERVICES

Most but not all dwellings and other structures are connected in some way to a wide range of utilities and services. In the majority of cities, towns and villages structures are connected to key utilities and services, such as a sewer system, potable (drinking) water, gas and electricity. This is not always the case for more remote structures, however.

Whenever construction work is carried out, whether it is on an existing structure or a new build, the supply of utilities and services or the linking up of these parts of the infrastructure are very important. Often they will require the services of specialist engineers from the service provider.

KEY TERMS

Infrastructure

– these are basic facilities, such as a power supply, a road network and a communication link.

Service provider

– these are companies or organisations that provide utilities, such as gas, water, communications or electricity.

Table 3.6 outlines the main utilities and services that are provided to most structures.

Utility or service	Description
Drainage	Drainage is delivered by a range of water and sewerage companies in the UK. They are responsible for ensuring that surface water can drain away into their system.
Waste water and sewerage	Any waste water and sewage generated by the occupants of a structure needs to have the necessary pipework to link it to the main sewerage system. It is then sent to a sewage treatment works via the pipework. If there is no connection to mains sewerage, the building may have a septic tank, which is a small-scale, self-contained sewage treatment system.
Water	Each structure should be linked to the water supply that provides wholesome, potable drinking water. The pipework linking the structure to the water supply needs to be protected to ensure that backflow from any other source does not contaminate the system.
Gas	Each area has a range of different gas suppliers. This is delivered via a service pipe from the main system into the structure. Areas that do not have access to the main gas supply system use gas contained in cylinders.
Electricity	The National Grid provides electricity to a variety of different electricity suppliers. It is the National Grid that operates and maintains the cabling. There are around 28 million individual electricity customers in the UK.
Communications (telephone, data, cable)	There are several ways in which telecommunications can be linked to a structure. Traditional telephone poles hold up copper cables and not only provide telephone but also internet access to structures. In cities and many of the larger towns poles are being replaced by cables that are fibre optic and run underground. These are then linked to each individual structure.
Ducting (heating and ventilation)	Heating and ventilation engineers install and maintain duct work. The complex systems are known as HVAC. These systems can transfer air for heating or cooling of the structure. The overall system can also provide hot and cold water systems, along with ventilation.

Table 3.6

KEY TERMS

HVAC

– this is an abbreviation for 'heating, ventilation and air-conditioning'. This has been a service provided to many industrial buildings for a number of years, but it is now becoming more common in domestic dwellings, particularly new developments.

SUSTAINABILITY AND INCORPORATING SUSTAINABILITY INTO CONSTRUCTION PROJECTS

Carbon is present in all fossil fuels, such as coal or natural gas. Burning fossil fuels releases carbon dioxide, which is a greenhouse gas linked to climate change.

Energy conservation aims to reduce the amount of carbon dioxide in the atmosphere. The idea is to do this by making buildings better insulated and, at the same time, make heating appliances more efficient. It also means attempting to generate energy using renewable and/or low or zero carbon methods.

According to the government's Environment Agency, sustainable construction is all about using resources in the most efficient way. It also means cutting down on waste on site and reducing the amount of materials that have to be disposed of and put into landfill.

In order to achieve sustainable construction the Environment Agency recommends:

* reducing construction, demolition and excavation waste that needs to go to landfill

* cutting back on carbon emissions from construction transport and machinery

* responsibly sourcing materials

* cutting back on the amount of water that is wasted

* making sure construction does not have an impact on biodiversity.

Sustainable construction and incorporating it into construction projects

Recently the idea of sustainable construction has focused on ensuring that the building is not only of good quality and affordable, but also that it is efficient.

Sustainable construction also means having the least negative environmental impact. So this means minimising the use of raw materials, energy, land and water. This is not only during the construction phase but also for the lifetime of the building.

Finite and renewable resources

We all know that resources such as coal and oil will eventually run out. These are examples of finite resources.

Oil is not just used as fuel – it is in plastic, dyes, lubricants and textiles. All of these are used in the construction process.

Renewable resources are those that are produced either by moving water, the sun or the wind. They include materials that come from plants, such as biodiesel, or the oils used to make adhesives.

The construction process itself is only part of the problem. It is also the longer term impact and demands that the building will have on the environment. This is why there has been a drive towards sustainable homes and there is a Code for Sustainable Homes (an environmental assessment method for rating and certifying the performance of new homes).

KEY TERMS

Landfill

– 170 million tonnes of waste from homes and businesses are generated in England and Wales each year. Much of this has to be taken to a site to be buried.

Biodiversity

– wherever there is construction there is a danger that wildlife and plants could be disturbed or destroyed. Protecting biodiversity ensures that at risk species are conserved.

Figure 3.22 Most modern new-builds follow sustainable principles

Construction and the environment

In 2010 construction, demolition and excavation produced 20 million tonnes of waste that had to go into landfill. The construction industry is also responsible for most illegal fly tipping (illegally dumping waste). In any year the Environment Agency responds to around 350 serious pollution incidents caused as a result of construction.

Regardless of the size of the construction job, everyone working on the project is responsible for the impact they have on the environment. Good site layout, planning and management can help reduce these problems.

Sustainable construction helps to encourage this because it means managing resources in a more efficient way, reducing waste and reducing your **carbon footprint.**

Architecture and design

The Code for Sustainable Homes Rating Scheme was introduced in 2007. Many local authorities have instructed their planning departments to encourage sustainable development. This begins with the work of the architect who designs the building.

Local authorities ask that architects and building designers:

* ensure the land is safe for development – that if it is contaminated this is dealt with first

* ensure there is access to and protection of the natural environment – this helps ensure biodiversity and tries to create open spaces for local people

* reduce the negative impact on the local environment – any buildings keep noise, air, light and water pollution down to a minimum

* conserve natural resources and cut back carbon emissions – this includes use of energy, materials and water during construction and the life of the building

* ensure comfort and security – good access, close to public transport, safe parking and protection against flooding.

Figure 3.23 Sustainable developments aim to be pleasant places to live

Using locally managed resources

The construction industry imports nearly 6 million cubic metres of sawn wood each year. However there is plenty of scope to use the many millions of cubic metres of timber produced in managed forests in the UK, particularly in Scotland.

Local timber can be used for a wide variety of different construction projects:

* softwood – including pines, firs, larch and spruce – for panels, decking, fencing and internal flooring

* hardwood – including oak, chestnut, ash, beech and sycamore – for a wide variety of internal joinery.

Using local materials reduces transportation costs and time, minimises the project's carbon footprint and means that there is less chance for the materials to be damaged in transit.

Eco-friendly, sustainable manufactured products and environmentally resourced timber

There are now many suppliers that offer sustainable building materials as a green alternative. Some tiles, for example, are now made from recycled plastic bottles and stone particles.

There is now a National Green Specification database of all environmentally friendly building materials. This provides a checklist where it is possible to compare specifications of sustainable products to traditionally manufactured products, such as bricks.

Simple changes can be made, such as using timber or ethylene-based plastics instead of UPVC window frames to ensure a building uses more sustainable materials.

As we have seen, finding locally managed resources, such as timber, makes sense in terms of cost and in terms of protecting the environment. There are always alternatives to the use of traditional resources that could affect the environment.

The Timber Trade Federation produces a timber certification system. This ensures that wood products are labelled to show that they are produced in sustainable forests.

Around 80 per cent of all the softwood used in construction comes from Scandinavia or Russia. Another 15 per cent comes from the rest of Europe, or even North America. The remaining 5 per cent comes from tropical countries, and is usually sourced from sustainable forests.

Alternative methods of building

The most common type of construction is, of course, brick and blockwork. However there are plenty of other options:

* timber frame – using green oak

* insulated concrete formwork – where a polystyrene mould is filled with reinforced concrete

Figure 3.24 Window frames made from timber

Figure 3.25 Timber Certification System

* structural insulated panels – where buildings are made up of rigid building boards rather like huge sandwiches

* modular construction – this uses similar materials and techniques to standard construction, but the units are built off site and transported ready-constructed to building site where they are connected together.

Figure 3.26 Green roofing

Figure 3.27 Flooring made from cork

DID YOU KNOW?

The Forest Stewardship Council has a system that verifies that wood comes from well-managed forests. The Programme for the Endorsement of Forest Certification Schemes promotes sustainable managed forests throughout the world (www.fsc.org and www.pefc.co.uk).

KEY TERMS

Biodegradable

– this material will more easily break down when it is no longer needed. This breaking down process is done by micro-organisms.

Organic

– these are natural substances, usually extracted from plants.

There are alternatives to traditional flooring and roofing, all of which are greener and more sustainable. Green roofing has become an increasing trend in recent years. Metal roofs made of steel, aluminium and copper use a high percentage of recycled material. Solar roof shingles, or solar roof laminates, while expensive, decrease the cost of electricity and related heating costs of the dwelling. Some buildings even have a waterproof membrane, which is covered with a growing medium and planted with vegetation like sedum plants. This provides additional insulation, absorbs air pollution, helps to collect and process rainwater and keeps the roof surface temperature down.

Just as roofs are becoming greener, so too are the options for flooring. The use of bamboo, eucalyptus and cork is becoming more common. A new version of linoleum has been developed with biodegradable, organic ingredients. Some buildings are also using sustainable alternatives to traditional timber floorboards and joists, and these can be coloured, stained or patterned.

An increasing trend has been for what is known as off-site manufacture (OSM). European OSM businesses, particularly those in Germany, have built over 100,000 houses. The entire house is manufactured in a factory and then assembled on site. Walls, floors, roofs, windows and doors with built-in electrics and plumbing, all arrive on a lorry. Some manufacturers even offer completely finished dwellings, including carpets and curtains. Many of these modular buildings are designed to be far more energy efficient than traditional brick and block constructions. Many come ready fitted with heat pumps, solar panels and triple-glazed windows.

Energy efficiency and incorporating it into construction projects

Energy efficiency is all about using less energy to provide the same level of output. Governments are working towards the world's energy needs by 30 per cent before 2050. This means producing more energy efficient buildings. It also means using energy efficient methods to produce materials and resources needed to construct buildings.

Building Regulations

In terms of energy conservation, the most important UK law is the Building Regulations 2010, particularly Part L. The Building Regulations:

* list the minimum efficiency requirements

* provide guidance on compliance, the main testing methods, installation and control

* cover both new dwellings and existing dwellings.

A key part of the regulations is the Standard Assessment Procedure (SAP), which measures or estimates the energy efficiency performance of buildings.

Local planning authorities also now require that all new developments generate at least 10 per cent of their energy from renewable sources. This means that each new project has to be assessed one at a time.

Energy conservation

By law, each local authority is required to reduce carbon dioxide emissions and to encourage the conservation of energy. This means that everyone has a responsibility in some way to conserve energy:

* Clients, along with building designers, are required to include energy efficient technology in the build.

* Contractors and sub-contractors have to follow these design guidelines. They also need to play a role in conserving energy and resources when working on site.

* Suppliers of products are required by law to provide information on energy in the production of their products.

In addition, new energy efficiency schemes and Building Regulations cover the energy performance of buildings. Each new build is required to have an Energy Performance Certificate. This rates a building's energy efficiency from A (which is very efficient) to G (which is very inefficient).

Some building designers have also begun to adopt other voluntary ways of attempting to protect the environment. These include BREEAM, which is an environmental assessment method, and the Code for Sustainable Homes, which is a certification of sustainability for new builds.

Figure 3.28 The Energy Saving Trust encourages builders to use less wasteful building techniques and more energy efficient construction

High, low and zero carbon

When we look at energy sources, we consider their environmental impact in terms of how much carbon dioxide they release. Accordingly, energy sources can be split into three different groups:

* high carbon – those that release a lot of carbon dioxide

* low carbon – those that release some carbon dioxide

* zero carbon – those that do not release any carbon dioxide.

Some examples of high carbon, low carbon and zero carbon energy sources are given in Table 3.7.

High carbon energy source	Description
Natural gas or LPG	Piped natural gas or liquid petroleum gas stored in bottles
Fuel oils	Domestic fuel oil, such as diesel
Solid fuels	Coal, coke and peat
Electricity	Generated from non-renewable sources, such as coal-fired power stations
Low carbon energy source	
Solar thermal	Panels used to capture energy from the sun to heat water
Solid fuel	Biomass such as logs, wood chips and pellets
Hydrogen fuel cells	Convert chemical energy into electrical energy
Heat pumps	Convert low temperature heat into higher temperature heat
Combined heat and power (CHP)	Generates electricity as well as heat for water and space heating
Combined cooling, heat and power (CCHP)	A variation on CHP that also provides a basic air conditioning system
Zero carbon energy	
Electricity/wind	Uses natural wind resources to generate electrical energy
Electricity/tidal	Uses wave power to generate electrical energy
Hydroelectric	Uses the natural flow of rivers and streams to generate electrical energy
Solar photovoltaic	Uses solar cells to convert light energy from the sun into electricity

Table 3.7

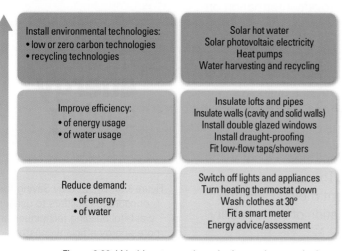

Figure 3.29 Working towards reducing carbon emissions

It is important to try to conserve non-renewable energy so that there will be sufficient fuel for the future. The idea is that finite sources of fuel should last as long as is necessary to completely replace it with renewable sources, such as wind or solar energy.

Alternative heating sources

There are several new ways in which we can harness the power of water, the sun and the wind to provide us with new heating sources. All of these systems are considered to be far more energy efficient than traditional heating systems, which rely on gas, oil, electricity or other fossil fuels.

Solar thermal

At the heart of this system is the solar collector, which is often referred to as a solar panel. The idea is that the collector absorbs the sun's energy, which is then converted into heat. This heat is then applied to the system's heat transfer fluid.

The system uses a differential temperature controller (DTC) that controls the system's circulating pump when solar energy is available and there is a demand for water to be heated.

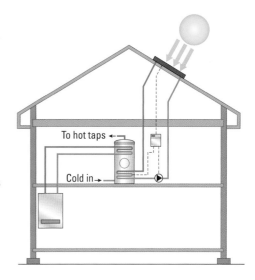

Figure 3.30 Solar thermal hot water system

In the UK, due to the lack of guaranteed solar energy, solar thermal hot water systems often have an auxiliary heat source, such as an immersion heater.

Biomass (solid fuel)

Biomass stoves burn either pellets or logs. Some have integrated hoppers that transfer pellets to the burner. Biomass boilers are available for pellets, woodchips or logs. Most of them have automated systems to clean the heat exchanger surfaces. They can provide heat for domestic hot water and space heating.

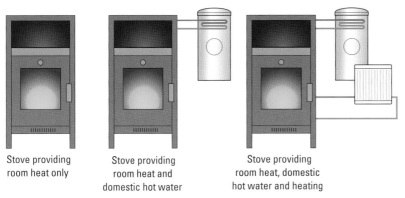

Stove providing
room heat only

Stove providing
room heat and
domestic hot water

Stove providing
room heat, domestic
hot water and heating

Figure 3.31 Biomass stoves output options

Heat pumps

Heat pumps convert low temperature heat from air, ground or water sources to higher temperature heat. They can be used in ducted air or piped water **heat sink** systems.

There are a variety of different arrangements for each of the three main systems:

* Air source pumps operate at temperatures down to minus 20°C. They have units that receive incoming air through an inlet duct.

* Ground source pumps operate on **geothermal** ground heat. They use a sealed circuit collector loop, which is buried either vertically or horizontally underground.

KEY TERMS

Heat sink

– this is a heat exchanger that transfers heat from one source into a fluid, such as in refrigeration, air-conditioning or the radiator in a car.

Geothermal

– relating to the internal heat energy of the earth.

* Water source systems can be used where there is a suitable water source, such as a pond or lake.

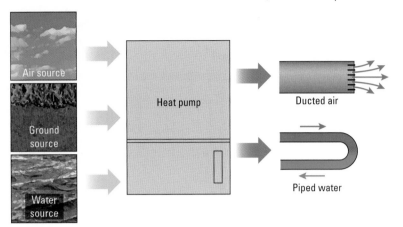

Figure 3.32 Heat pump input and output options

The heat pump system's efficiency relies on the temperature difference between the heat source and the heat sink. Special tank hot water cylinders are part of the system, giving a large surface-to-surface contact between the heating circuit water and the stored domestic hot water.

Combined heat and power (CHP) and combined cooling heat and power (CCHP) units

These are similar to heating system boilers, but they generate electricity as well as heat for hot water or space heating (or cooling). The heart of the system is an engine or gas turbine. The gas burner provides heat to the engine when there is a demand for heat. Electricity is generated along with sufficient energy to heat water and to provide space heating.

CCHP systems also incorporate the facility to cool spaces when necessary.

Wind turbines

Freestanding or building-mounted wind turbines capture the energy from wind to generate electrical energy. The wind passes across rotor blades of a turbine, which causes the hub to turn. The hub is connected by a shaft to a gearbox. This increases the speed of rotation. A high speed shaft is then connected to a generator that produces the electricity.

Solar photovoltaic systems

A solar photovoltaic system uses solar cells to convert light energy from the sun into electricity. The solar cells are usually made of silicon and are semi-conductors. The sunlight hits the solar cells and photons are absorbed. This causes negatively charged electrons in the cell to detach from their atoms and flow through the cell to create electricity. The electricity is direct current (dc). The dc current is then converted by an inverter to alternating current (ac), which is the type of current used for mains electricity.

Energy ratings

Energy rating tables are used to measure the overall efficiency of a dwelling, with rating A being the most energy efficient and rating G the least energy efficient.

Alongside this is an environmental impact rating (see Fig 3.38). This measures the dwelling's impact on the environment in terms of how much carbon dioxide it produces. Again, rating A is the highest, showing it has the least impact on the environment, and rating G is the lowest.

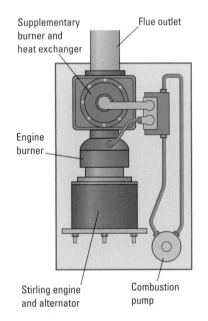

Figure 3.33 Example of a MCHP (micro combined heat and power) unit

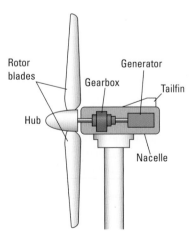

Figure 3.34 A basic horizontal axis wind turbine

A standard assessment procedure (SAP) is used to place the dwelling on the energy rating table. This will take into account:

* the date of construction, the type of construction and the location

* the heating system

* insulation (including cavity wall)

* double glazing.

The ratings are used by local authorities and other groups to assess the energy efficiency of new and old housing, and must be provided to potential purchasers when houses are sold.

Preventing heat loss

Most old buildings are under-insulated and would benefit from additional insulation, whether this is ceilings, walls or floors.

The measurement of heat loss in a building is known as the U Value. It measures how well parts of the building transfer heat. Low U Values represent high levels of insulation. U Values are becoming more important as they form the basis of energy and carbon reduction standards.

By 2016 all new housing is expected to be Net Zero Carbon. This means that the building should not be contributing to climate change.

Many of the guidelines are now part of Building Regulations (Part L). They cover:

* insulation requirements

* openings, such as doors and windows

* solar heating and other heating

* ventilation and air conditioning

* space heating controls

* lighting efficiency

* air tightness.

Building design

UK households spend £2.4bn every year just on lighting. One of the ways of tackling this cost is to use energy saving lights, but also to maximise natural lighting. For the construction industry this means:

* increased window size

* orientating building angles to make the most of sunlight – south facing windows maximise sunlight in winter and limit overheating in the summer

* considering window design by using windows with a variety of different types of opening to allow ventilation.

Solar tubes are another way of increasing light. These are small domes on the roof, which collect sunlight and then direct it through a tube (which is reflective). It is then directed through a diffuser in the ceiling to spread light into the room.

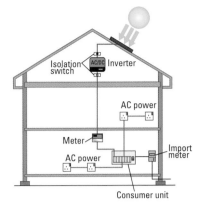

Figure 3.35 A basic solar photovoltaic system

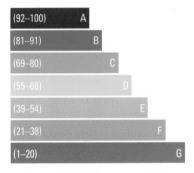

Figure 3.36 SAP energy efficiency rating table. The ranges in brackets show the percentage energy efficiency for each banding

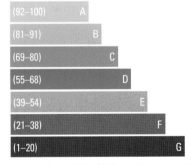

Figure 3.37 SAP environmental impact rating table

TEST YOURSELF

1. In which of the following types of buildings is a traditional strip foundation used?

    a. High rise

    b. Medium rise

    c. Low rise

    d. Industrial buildings

2. Which of the following is a reason for using a raft foundation?

    a. The subsoil is rock

    b. The subsoil is unstable

    c. The subsoil is stable

    d. The access to the site allows it

3. What holds down a floating floor?

    a. Nails and screws

    b. Adhesives

    c. Blocks

    d. Its own weight

4. What is another term for formwork?

    a. Shuttering

    b. Cavity

    c. Joist

    d. Boarding

5. What is the minimum distance the DPC should be above ground level?

    a. 50 mm

    b. 100 mm

    c. 150 mm

    d. 200 mm

6. A roof is said to be flat if it has a slope of less than how many degrees?

    a. 5

    b. 10

    c. 15

    d. 20

7. What shape is the upper part of a gable end?

    a. Rectangular

    b. Semi-circular

    c. Square

    d. Triangular

8. What do you call the horizontal timber that is placed at the top of a wall at eaves level in a roof, to hold the ends of joists or rafters?

    a. Fascia

    b. Bracings

    c. Wall plate

    d. Batten

9. What happens to the majority of construction demolition and excavation waste?

    a. It is buried on site

    b. It is burned

    c. It goes into landfill

    d. It is recycled

10. Which part of the Building Regulations 2010 requires the construction industry to consider and use energy efficiently?

    a. Part B

    b. Part D

    c. Part K

    d. Part L

Chapter 4

Unit CSA–L1Occ09
PRODUCE WOODWORKING JOINTS

LEARNING OUTCOMES

LO1/2: Know how to and be able to prepare resources for producing woodworking joints

LO3: Know woodworking materials and their storage requirements

LO4/5: Know how to and be able to produce woodworking joints

INTRODUCTION

The aims of this chapter are to:

* show you how to form woodworking joints

* help you to select and mark out materials

* help you to select and use hand tools.

PREPARING RESOURCES FOR PRODUCING WOODWORKING JOINTS

When you are producing woodworking joints it is important that you begin by selecting the right materials and tools. Accurate marking out following either full-size or scaled drawings is also very important.

Like all carpentry jobs you also need to be aware of health and safety risks and wear the appropriate PPE. As we will see, it is also important to know about the structure of wood and how it can be worked.

Hazards, health and safety and risk assessment

Probably the greatest hazards from hand tools are as a result of them being misused or poorly maintained. Tools such as chisels can be dangerous to use because they can slip if they are hit with a hammer or mallet. In extreme cases of misuse the handle or blade could snap.

It is your responsibility and that of your employer to make sure that any tools you are using are in good condition and safe to be used. The main causes of injury are:

* not using the right tool for the job

* not having been trained to use the tool in a safe way

* deliberately using the wrong tool

* failing to maintain the tool

* not wearing the right PPE.

PPE

You should always wear eye protection when producing woodworking joints. You are likely to create dust and fragments of wood. These are potentially very dangerous, particularly to the eyes.

You should always wear some kind of hand protection too. The thickness of the gloves will depend on the job, as you may need to be able to hold things, which can be difficult if your gloves are too thick. It may seem inconvenient and hinder your work but suitable gloves will protect your hands against cuts, abrasions and most impacts.

More information about suitable PPE and the laws concerning it can be found in Chapter 1.

PRACTICAL TIP

It is the responsibility of everyone on site to ensure that appropriate PPE is worn.

Using tools safely

Tools need to be regularly inspected and will need routine maintenance to stop them from becoming potential hazards. Cutting tools, such as wood chisels, will need to be regularly sharpened. Wooden handles must be free of cracks and splinters.

If a tool cannot be properly repaired it should be replaced.

There are many different ways of controlling hazards:

* Pick the right tool for the job – it should be comfortable to use.

* Use a vice, clamp or bench holdfast to hold the timber in place while you work on it.

* Make sure that you are in a comfortable position when you are working and do not overstretch yourself.

Remember that poorly maintained hand tools are more dangerous. Tradespeople use quality tools because they will last a lifetime if used and maintained correctly. If tools are of inferior quality, they are harder to maintain, which increases the risk of personal injury.

Working drawings and marking out

Working drawings are either full size or scaled, accurate illustrations of the kind of joint that you are going to be producing. From these working drawings you will be able to mark off your pieces of timber to the required dimensions to allow you to make the necessary cuts. The shape and size of the final piece will also help you decide which tools you will need to achieve the task.

Once you have a set of working drawings for the jobs you will be able to prepare a **cutting list.**

There are a number of tools for measuring and marking out. You can use these to make all the necessary pencil marks on the timber before you start cutting and shaping. These are shown in Table 4.1.

KEY TERMS

Cutting list

– this is an itemised list, with quantities, length, width, thickness and material needed to complete a job. It also notes exactly what the piece of wood will be needed for.

Measurement or marking tool	Description
Callipers and dividers 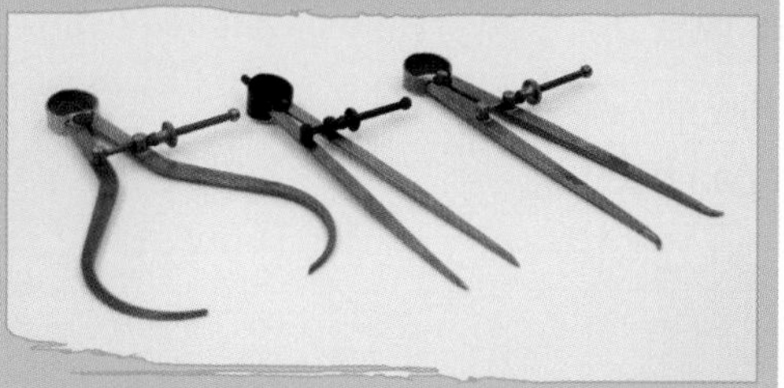 Figure 4.1	These are used to set out measurements.
Combination square  Figure 4.2	As its name suggests, a combination square can be used for a number of operations, including marking, squaring and measuring.
Cutting gauge  Figure 4.3	This is used for marking lines into timber across the grain. It has a knife-shaped blade that makes the cut. It is often used in the marking out of dovetails.
Spirit level  Figure 4.4	Spirit levels come in various sizes. They are for checking plumb (vertical) and level (horizontal).
Marking gauge  Figure 4.5	These are used for marking lines parallel to the edge of a piece of timber. They use a spur or point to make the mark.

Measurement or marking tool	Description
Marking knife  Figure 4.6	This is an alternative to marking out with a pencil. They are usually used on hardwoods and are considered to be more accurate.
Mitre square  Figure 4.7	This is a metal blade that is fixed to a wooden stock at 45°. It is used for marking out mitre cuts.
Mortise gauge  Figure 4.8	This is used to set out mortise and tenon joints. One of the spurs is fixed and the other is adjustable.
Pencils 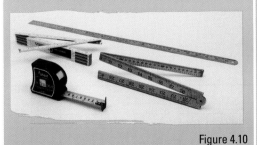 Figure 4.9	These are graded according to hardness or softness of the lead. Carpenters will use a carpenter's pencil on carcassing work and a 4H pencil when marking out joinery. They should be kept sharp at all times.
Rules and measures Figure 4.10	Various options are available. Standard steel measuring tapes, folding rules and metal steel rules are all used.

Measurement or marking tool	Description
Sliding bevel 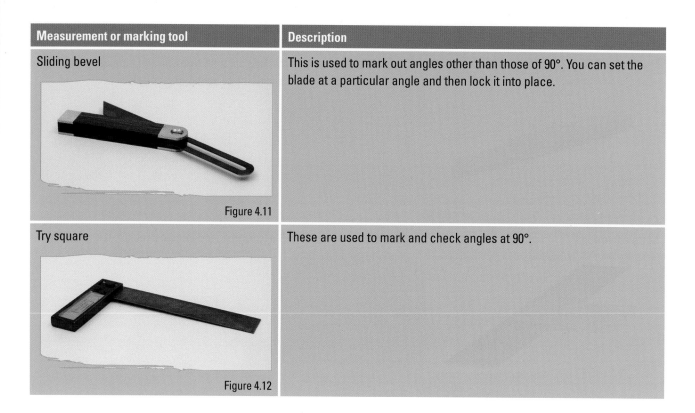 Figure 4.11	This is used to mark out angles other than those of 90°. You can set the blade at a particular angle and then lock it into place.
Try square Figure 4.12	These are used to mark and check angles at 90°.

Table 4.1

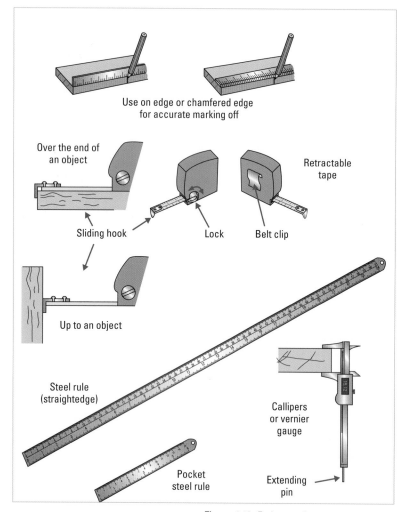

Figure 4.13 Rules and tape measures

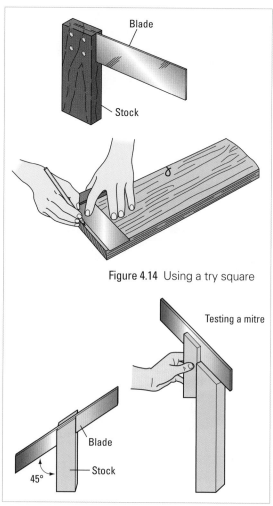

Figure 4.14 Using a try square

Figure 4.15 Mitre square

Figure 4.16 Combination square

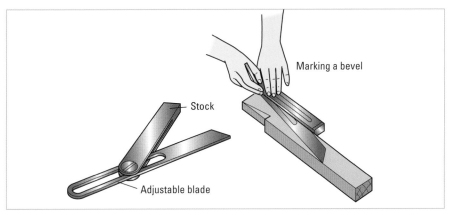

Marking a bevel

Stock

Adjustable blade

Figure 4.17 A sliding bevel

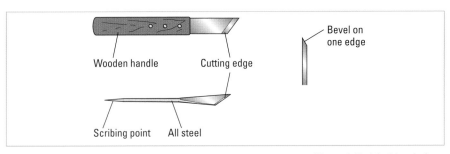

Bevel on one edge

Wooden handle   Cutting edge

Scribing point   All steel

Figure 4.18 Marking knives

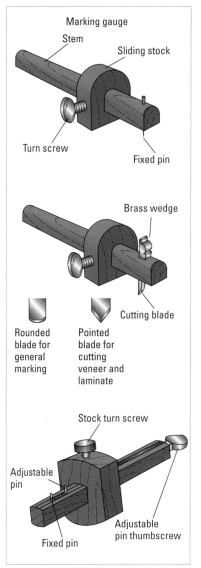

Marking gauge

Stem

Sliding stock

Turn screw

Fixed pin

Brass wedge

Cutting blade

Rounded blade for general marking

Pointed blade for cutting veneer and laminate

Stock turn screw

Adjustable pin

Fixed pin

Adjustable pin thumbscrew

Figure 4.19 Carpenter's gauges

Woodworking hand tools

The focus of this chapter is about choosing the right tools for the particular woodworking joint you are producing.

For the purposes of this unit you need to know the uses of:

* the range of different chisels

* mallets

* a range of hand saws

* hammers, including claw hammers and cross pein or Warrington hammer types

* planes

* wood boring tools – drills and drill bits

* squares

* a range of types of screwdrivers

* gauges and measures.

Routine maintenance of hand tools

Tools such as chisels, hand saws and planes need to be checked on a regular basis to make sure that they are still in good working order. This means that maintenance of tools with sharp cutting edges should be a regular part of your routine. In Chapter 5 there is advice on how to sharpen hand tools, either using sharpening stones or grinding wheels.

Other tools can also become worn over time. Mallets and hammers need to have solid and split-free handles and heads. The drill bits of wood boring tools will eventually dull and become less effective. Screwdrivers can also gradually degrade, making it difficult to drive screws into or get screws out of timber. A Phillips or Pozidriv screwdriver should be replaced if the end is unusable; however, a slotted screwdriver blade can be reground.

It does not matter how expert you become in producing woodworking joints, if your tools are not correctly maintained and sharpened, it will be difficult to achieve the quality required.

Reporting and recording faulty or defective equipment

Carpenters use their own hand tools, so it is their responsibility to ensure they are of sufficient quality. Your employer or supervisor will expect you to maintain and sharpen your hand tools. It is good practice to keep your tools well maintained at all times.

Equipment other than the carpenter's hand tools will be the responsibility of both employer and employee. A regular maintenance plan should be in place, which should include reporting faults and a method of recording maintenance and repair.

If your employer or college has provided tools that you find to be faulty or defective, you should report this to your supervisor or tutor. They will make a judgement on whether the equipment needs to be replaced or repaired.

The tools required for marking out timber

Usually to mark out you will need:

* a measure or rule

* a square

* marking gauge

* pencil.

Ensuring the accuracy of marking out tools

Before any marking out can be done, the timber should be dimensionally accurate and fit for purpose.

In order to make sure that your marking out tools are accurate you should do the following:

* Check a tape against a steel rule that you know to be accurate.

* For squares you should mark a line on a piece of waste timber and then reverse the square to check to see whether the line is straight.

* For gauges you should ensure the spurs are sharp and straight.

* Ensure pencils are sharp and the correct grade.

CASE STUDY

South Tyneside Homes

South Tyneside Council's Housing Company

Develop your hand skills first

Glen Campbell is a Property Services Team Leader at South Tyneside Homes. He finished his apprenticeship in carpentry and joinery in 1997.

'When I was interviewed for my apprenticeship at South Tyneside Homes, they told me how I would go to college for three years on block release and then day release. It was in those first six weeks at college where you learnt all your basic hand skills.

The biggest tip I can give to new learners is to develop your hand skills first. Before you move on to machines at all, you have to practise with your hand tools, chisels, hand saws, mallets, hammers. It does take time, but you should be comfortable with everything, from taking a piece of timber down from rough to dressed, to creating a square edge on a piece of timber. That was one of the first things we learnt to do.

Setting your hand plane to get that angle is not as easy as you think! You could see people getting frustrated when they'd set the square on and it'd have a little bubble, which they'd then have to take off. But it felt great when you'd mastered it because to move on, you'd have to prove that you could do that first. You can only progress once you get that level of detail right. In the workplace, you'll be working to high-level specifications, using real materials. "Measure twice, cut once" – it's an old saying, but it really does apply.'

WOODWORKING MATERIALS AND THEIR STORAGE REQUIREMENTS

Conversion

Different methods used for timber conversion

Timber conversion describes the way in which the tree logs are transformed into boards or planks. The way in which they are cut has a direct impact on what you can use the wood for while on site or in the workshop. There are various common ways that timber can be converted.

Wood can be cut from the tree trunk in one of two different ways:

* tangential –when the wood is cut in such a way as that the annual growth rings are on the widest surface of the wood at around half its width and at an angle of less than 45°

* radial – the annual rings are again at the widest surface of the finished piece of wood, at an angle of 45° or more.

Through and through

This is one of the oldest ways in which wood is sawn. It is still one of the simplest and produces very little waste although the boards may distort easily. The boards that come from the outside of the trunk are tangential. These account for around 60 per cent of the boards. The remainder are radial.

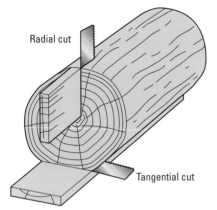
Radial cut
Tangential cut

Figure 4.20 Timber conversion

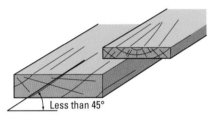
Less than 45°

Figure 4.21 Tangential

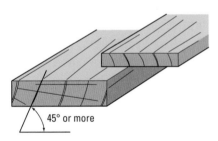
45° or more

Figure 4.22 Radial

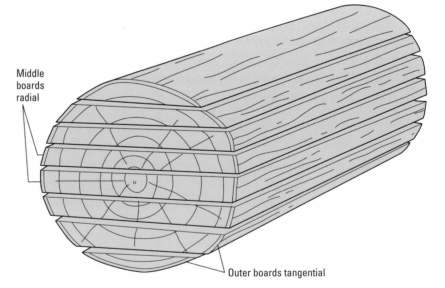
Middle boards radial

Outer boards tangential

Figure 4.23 Through and through (slab sawn)

Quarter sawn

Quarter sawing aims to produce top quality wood that can be used for joinery. It is one of the most expensive ways of producing wood. The boards are radially cut so that they are less likely to distort or shrink than other types of board. In some hardwoods, the cut will reveal the figure, which is regarded as both attractive and decorative.

Medullary rays are groups of food storage cells and look as if they radiate from the pith outwards towards the bark. They can be cut and planed to show the rays off as a decorative feature.

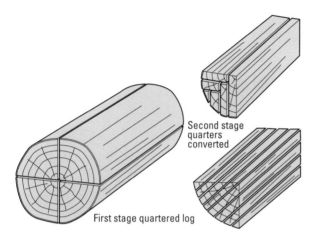

Second stage
quarters
converted

First stage quartered log

Figure 4.24 Quarter sawn

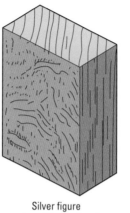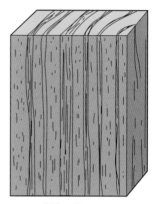

Silver figure
in species with prominent rays

Ribbon figure
in species with interlocking grain

Figure 4.25 Decorative figure in quarter sawn timber

Tangential

Tangential cutting is chosen because it produces extremely strong material. It also produces very decorative wood, which is known as flame figuring.

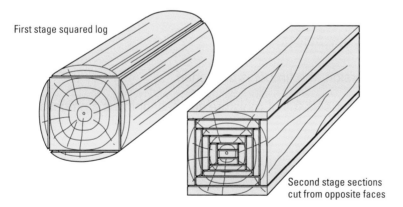

First stage squared log

Second stage sections
cut from opposite faces

Figure 4.26 Tangential sawn

Boxed heart

Boxed heart conversion is a method of converting the timber to use as much of the log as possible, especially when there is visible damage to the heart, or centre, of the trunk. This may be because the tree suffered from shock when it was young or it had begun to rot away. Alternatively, the boxed heart method can be used for structural timbers. If the heart or pith is in good condition, when the bark is removed the timber will retain its strength. If the centre of the tree shows decay, this will need to be removed. The following shows two methods of doing this:

* It can be quarter sawn and the good wood used for flooring.

* It can be tangentially cut, which will provide more wood and much less waste.

The boxed heart is often used for structural framing.

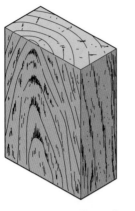

Figure 4.27 Flame figuring

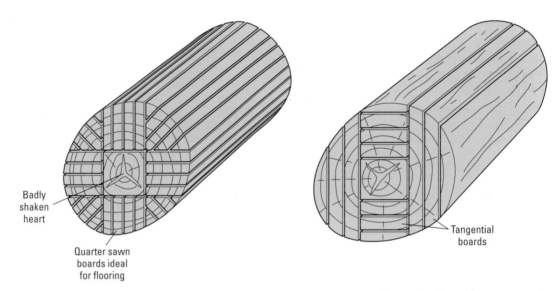

Badly shaken heart

Quarter sawn boards ideal for flooring

Tangential boards

Figure 4.28 Boxed heart conversion

Seasoning timber

After conversion, the timber is seasoned.

Newly cut timber is full of moisture and seasoning is a process used to dry out the wood.

The purpose of seasoning timber is to:

* reduce its moisture content. Timber may get dry rot if the moisture content is 20% or higher

* make timber more workable with tools

* remove the sap, as this will putrefy (decay) if it is left within the wood

* allow the wood to shrink to its dried-out size before the wood is used

* make the wood less likely to split or distort.

Methods of seasoning

The three different ways of seasoning wood are naturally allowing the wood to dry, which is called air seasoning or air drying; an artificial drying process known as kiln seasoning or drying; or water seasoning and drying.

When the wood is being air seasoned it is usually stacked in a covered, open-sided shed. This is raised off the ground to prevent moisture from the ground getting into the timber. The wood is stacked to allow air to circulate. A typical timber stack or shed can be seen in Fig 4.29.

In this process the boards are laid so that the largest boards are at the bottom and the smallest are at the top. The boards are all laid horizontally. Each layer of boards is separated by piling sticks. These support the made boards and allow the air to circulate. The piling sticks are usually made of the same type of wood as the wood that is being dried, to stop staining. Usually the ends of the boards are painted or covered so that they do not dry out too quickly. This process is used for both hard and softwoods.

Another air seasoning method is where the boards are often stacked in the same order in which they were cut from the original log. This is known as a boule, which can be seen in Fig 4.30.

The other alternative is to kiln season or dry the wood. The wood is usually air seasoned for a time before it is stacked into the kiln. Once the wood is in the kiln, depending on the type of wood and the size, it will need to be left in there for anything from a few days to a month and a half.

There are two different types of kiln:

* **A compartment kiln** is made of either brick or concrete. The timber is stacked in the same way as for air seasoning. The air circulation is controlled by a fan. Steam sprays are used to control the moisture content, which stops the wood from drying too quickly. Heat is supplied by heating coils.

* **A progressive kiln** is much larger in terms of scale. Large trucks or pallets of wood are stacked and progressively moved deeper into the kiln. The timber is gradually dried using steam and heat. It tends to be used for processing the same species of wood cut to the same dimensions.

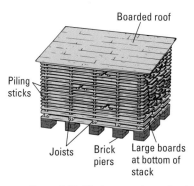

Figure 4.29 Timber stack or shed

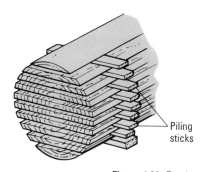

Figure 4.30 Boule

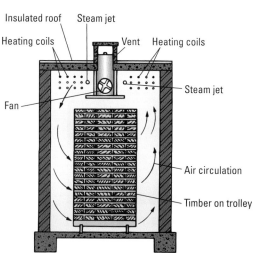

Figure 4.31 Compartment kiln

The final method is water seasoning, where newly cut timber is completely submerged in water for several weeks until most of the sap has washed away. When it is taken out of the water, it is laid out and left in freely circulating air to dry, being turned each day. Water seasoned timber warps and cracks less often than air seasoned or kiln seasoned timber but it can be brittle.

Figure 4.32 Timber waiting to be loaded into a progressive kiln

After seasoning, the timber is converted to commercial stock sizes.

Rough cut

Also known as sawn (or unwrot) rough cut wood is rough in texture. It follows normal sizing standards and is cut into boards. Timber that is rough sawn still has a precise measurement.

Figure 4.33 Rough cut wood

Planed

Planed all round (PAR) means the softwood or hardwood timber has a planed finish rather than a sawn finish. It is also known as PSE (planed sides and edges).

The end result will also vary for other resons:

* **The origin of the timber and its sustainability** – which part of the world the wood comes from and whether it was grown in a properly managed forest or plantation.

* **Common sizes** – there is a range of different shapes and sizes.

Figure 4.34 Planed all round

* **Sawn or planed** is whether the timber has been simply cut to a nominal sectional size or whether it is ready for immediate final use, having been planed to size. Sawn sizes are known as nominal and planed sizes are known as 'finished'. The timber's final use determines whether it will be used in its sawn state or in a planed form. For example floor joists are prepared to a given size so the only preparation they require is being sawn. Floorboards are an example of where additional preparation would be required because they need to be planed to finished dimensions and tonguing and grooving applied.

Storing woodworking materials securely and safely

The problem with all types of timber is that even if it has been seasoned and the moisture content reduced, if it is exposed to the elements then it will absorb more moisture. If the timber is allowed to get wet then dry out and get wet again, each time it dries out defects, faults and damage are likely to be caused.

Wood should not be stored in unsuitable conditions for any length of time. Even the strongest and largest pieces of timber will eventually be damaged. It should always be kept under cover. This prevents it from weathering (discolouring) or getting wet and then drying out (changes in moisture content).

Fig 4.35 shows different ways of storing timber. You should bear in mind the following:

* Large carcassing timbers and timber that will be used for external joinery should be kept off the ground. You should use piling sticks or stickers between each layer and the whole stack should be covered with a tarpaulin. This will protect it from rainwater and sunlight and should also allow air to circulate around the timbers.

* Internal joinery needs to be kept in a dry and well-ventilated store. Piling sticks should also be used if necessary.

PRACTICAL TIP

When you are working on a new building, it should be dried out after plaster and other 'wet' materials have been used. If this doesn't happen then kiln dried timber is likely to absorb some of the moisture. Often just ventilating the building is enough, although contractors may bring in warm air blowers to speed up the process.

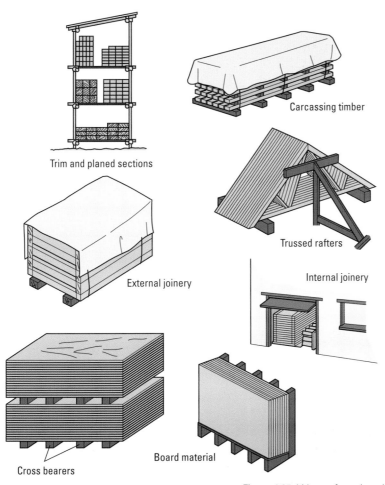

Trim and planed sections

Carcassing timber

Trussed rafters

External joinery

Internal joinery

Cross bearers

Board material

Figure 4.35 Ways of storing timber

Using the correct materials for the task

In this part of the chapter we look at the different sorts of wood and their properties, how they should be stored and how to identify common defects in wood. There are various different types of wood, which are known by their species. Each of these has different properties. The type of materials you use, for example softwoods or timber manufactured board, depends on where you work and the sort of work your company undertakes. Temperate wood is from trees that grow in the seasonal regions like Europe or North America. Most of their growth happens in spring and summer. Tropical wood is from trees that grow in hot climates, where it is warm enough for them to grow almost continuously, so are less likely to have seasonal 'growth rings', like temperate woods. Table 4.2 outline the main types of hardwoods and softwoods that you could come across.

Hardwoods

Name of wood	Type of wood	Description and uses
Ash  Figure 4.36	Temperate	This is a tough and flexible wood that has straight grain. It is often used for furniture and tool handles. Most of it comes from Europe.
Beech  Figure 4.37	Temperate	This has a fine texture, with hard and close grain. It is used to make furniture and woodblock floors and mainly comes from Europe.
Mahogany  Figure 4.38	Tropical	This can come from either Africa or South America. It is resistant to decay. It is used for high quality joinery, as well as veneers.

Name of wood	Type of wood	Description and uses
Maple Figure 4.39	Temperate	This is a wood from North America. It is fine grained and reasonably durable. It can be used for furniture, panelling and flooring.
Oak  Figure 4.40	Temperate	This is strong timber from temperate forests in the Northern Hemisphere. It is used for joinery, panelling and doors. Oak is both decorative and strong. It is used in furniture-making, structural framing, joinery and for gates and fencing.
Sapele  Figure 4.41	Tropical	This is from West Africa and is mainly used for furniture and veneers.
Teak Figure 4.42	Tropical	This is from South East Asia. It is used for joinery and is particularly suitable for exterior use as it contains its own natural preservatives. It is strong and durable.
Walnut Figure 4.43	Temperate	This is from European forests and is mainly used for veneers. It is quite strong and can be highly polished. It is decorative when in the form of a burr.

Softwoods

Name of wood	Type of wood	Description and uses
Douglas fir  Figure 4.44	North America	This wood has straight grain and is one of the hardest softwoods. It is used for a wide variety of joinery and is good for heavy structural tasks.
Pitch pine  Figure 4.45	North America	This very durable wood is used for joinery. It has similar properties to Douglas fir, although it is harder to work with due to its high resin content.
Redwood Figure 4.46	Europe	This wood is better known as pine. It is quite strong and can be used for a wide variety of internal and external work.
Western red cedar Figure 4.47	North America	This wood is extremely durable and is ideal for external use such as cladding, shingles and in conservatories. It has straight grain and does not need to be treated due to its naturally occurring oils.
Whitewood  Figure 4.48	Europe	This wood is also known as spruce. It has many similar properties to redwood in terms of its durability and strength.

Table 4.2

You will not always be working with solid wood, but in many cases you will be using timber manufactured boards. There is an increasing variety of these different types of board, which are detailed in Table 4.3.

Timber manufactured boards

Type of timber manufactured board	Description
Chipboard  Figure 4.49	This is made from a combination of flakes of wood and woodchips that have been compressed together and bonded using resin glue. It is available in many different grades, sizes and widths. It is used to make tongue and groove flooring.
Plywood  Figure 4.50	This is made of layers of timber, which are known as veneers. The veneers are glued together with alternating grain, which gives the plywood a good deal of strength and stability. Plywood is graded, which helps identify where different types can be used. Each side is graded separately so you could have a high grade on one side and a lower grade on the other. If it is marked with the stamp INT then it is only to be used for interior work. This is because the plywood is not resistant to humidity or dampness. If it is marked MR it means it has medium resistance to humidity and dampness. The best external type of plywood is marked WBP, which is weather and boil proof. However, WBP ply can delaminate when subjected to prolonged periods of exposure to wet weather conditions so for boats marine plywood is used. This must be compliant with BS 1088 and is made to perform better in wet and humid conditions, and is resistant to fungal attack. Fire-retardant, moisture-resistant and pressure-resistant plywood is available, which may have been treated with chemicals to give it the required properties.
Medium density fibreboard (MDF) 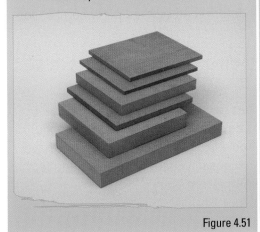 Figure 4.51	This fibreboard is made using a dry process. The fibres are bonded together with a resin adhesive. It is available in various thicknesses and sheet sizes. It is also available as ready-moulded skirting and mouldings. It is very easy to work with but should only be used internally. There are moisture-resistant versions available.

Type of timber manufactured board	Description
Block board 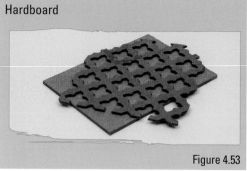 Figure 4.52	This is a form of laminated board. Strips of wood are laminated or stuck together and then veneers are glued to form a top and bottom surface. Usually the more strips inside the 'sandwich' the higher the quality of the laminated board. Block board has strips of up to 25 mm in width. Laminboard, which is a form of block board, has strips that are less than 25 mm thick. Block board is rarely used these days.
Hardboard Figure 4.53	This is made from wet wood fibres being compacted at a high temperature and pressure. This is available in a number of finishes, some of which are shaped and in thicknesses of up to 6 mm. There is an oil-tempered version of hardboard, which has some moisture resistance. Hardboard is used for numerous purposes in construction. It is extensively used in exhibition work as it is easily formed around curves.

Table 4.3

Properties of timbers

Timbers have different properties but may share some of the same characteristics. These properties are used to grade and class the timber. You should be aware of these as it may affect the wood you choose to use. These possible properties are as follows:

* **Workability** is how easy timber is to saw, plane and fix. It is graded good, medium and difficult.

* **Durability** is the timber's resistance to fungal decay. Class 1 is very durable (VD), Class 2 is durable (D), Class 3 is moderately durable (MD), Class 4 is slightly durable (SD) and Class 5 is not durable (ND). Grades and classes are determined by third party certification companies.

* **Cell structure** is different for softwood and hardwood (see details below).

* **The way in which the tree has grown** varies between species but again there is a difference between softwood and hardwood. Generally, softwoods have needle-like leaves and hardwoods have broader leaves. They are not necessarily 'hard' or 'soft' as the terms refer to the cellular structure of the tree.

Tables 4.4–4.6 summarise some of the key properties of commonly used timbers. The lower the number of the durability class, the more durable the wood.

Tropical hardwoods

Name of wood	Durability Class	Workability	Characteristics
Mahogany	2–3	Medium to good	Interlocking grain with a medium texture
Sapele	3	Medium	Interlocking grain with a medium texture
Teak	1	Medium	Straight grained (sometimes wavy) with medium texture

Table 4.4

Temperate hardwoods

Name of wood	Durability Class	Workability	Characteristics
Ash	5	Good	Straight grained with a coarse texture
Beech	5	Good	Straight grained with a fine texture
Maple	4	Medium	Straight grained with a fine texture
Oak (European)	2	Medium to difficult	Straight grained with a medium to coarse texture
Walnut	4	Medium	Interlocking grain with a medium texture

Table 4.5

Softwoods

Name of wood	Durability Class	Workability	Characteristics
Douglas fir	3	Good	Straight grained with a medium texture
Pitch pine	3	Medium	Straight grained with a medium texture
Redwood	4	Medium	Straight grained with a medium texture
Western red cedar	2	Good	Straight grained with a coarse texture
Whitewood	4	Good	Straight grained with a fine texture

Table 4.6

Cell structure

There are differences in the cell structure between softwoods and hardwoods. Softwoods have the simplest structure and only have two different types of cell:

* **Tracheids** are box-like cells. They are the main structural tissue and give the tree its strength. The sap rises from one of these cells to another.

* **Parenchyma** are the food storing cells that are found in the centre of the tree.

The illustrations below show a cube of softwood and the difference between fast and slow growing softwood. The slow-growing timber has thicker walls. This is because the tracheids are more numerous in slow-growing timber.

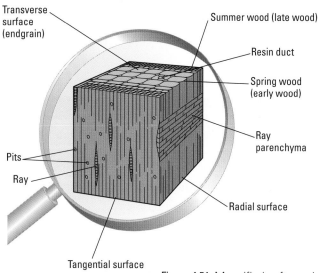

Figure 4.54 Magnified softwood cube

Figure 4.55 Slow and fast grown softwood

The cell structure of hardwood trees is more complicated. They have three different types of cell:

* **Fibres** are the cells that give the wood its strength. They are the main structural tissue.

* **Parenchyma** are the sap cells, which are either oval or circular and are food storing cells.

* **Vessels or pores** are the cells that allow the sap to move around to where it is needed in the tree.

Slow grown hardwoods have less fibre between their pores and are therefore weaker than fast grown hardwoods. A magnified view of a hardwood cube can be seen below, along with the difference between weak, slow grown hardwood and strong, fast grown hardwood. However, bear in mind that this is only a general rule of thumb – oak, for example, is a very strong, slow grown hardwood.

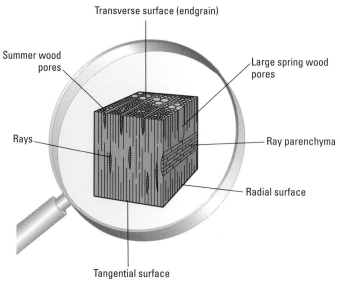

Figure 4.57 Magnified hardwood cube

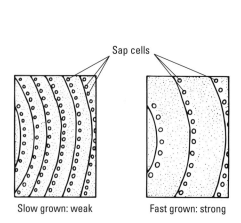

Figure 4.56 Fast and slow grown hardwood

Common defects found in timbers

Timber is a natural product so is unlikely to be perfect. Usually there are two reasons for a defect or imperfection in the wood:

* It may be naturally occurring, such as a knot.

* It may be caused through poor handling or seasoning of the wood after it has been cut.

Cupping
Cupping is when the wood is warped across the width of the board. It tends to be caused by poor seasoning.

Winding
Winding presents itself as a twisted board. This is often caused by the wood not being cut parallel to the pith (central core) of the tree.

Case hardening
Case hardening leaves the inner core of the wood still wet and the outer part of the wood dry. This is usually caused by the wood being dried too quickly. It can make machining the wood dangerous.

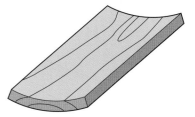

Figure 4.58 Cupping

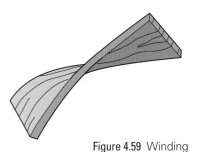

Figure 4.59 Winding

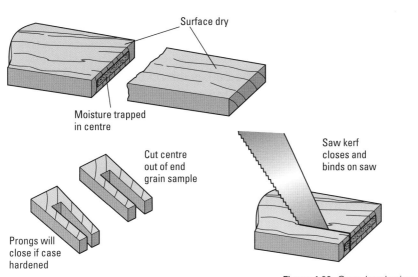

Surface dry

Moisture trapped in centre

Cut centre out of end grain sample

Saw kerf closes and binds on saw

Prongs will close if case hardened

Figure 4.60 Case hardening

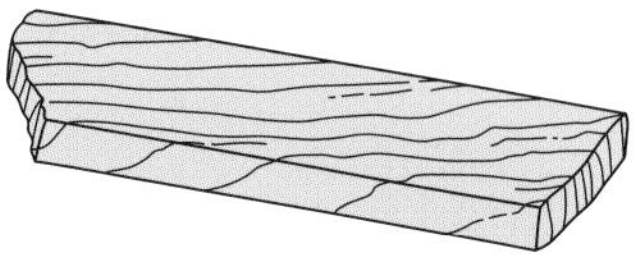

Figure 4.61 Sloping grain

Sloping grain
If the wood was not cut parallel to the pith of the tree then sloping grain can be seen. These reduce the bending strength of the timber.

Bowing
Bowing is also caused by poor seasoning and shows itself as bent along the length of the board.

Shakes
Shakes are naturally occurring defects, usually found in uncut logs and machined timber. There are a number of different types of shake that are named after the shape in which they appear. The shakes are splits between the annual rings, because tension has built in the tree while it was growing. If the wood is not seasoned properly and has dried out too quickly then a shake shape will appear.

Figure 4.62 Bowing

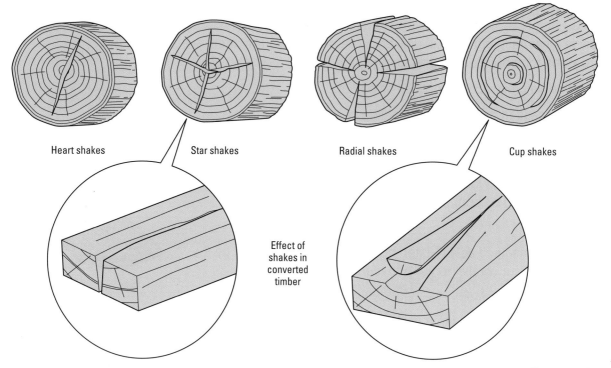

Heart shakes   Star shakes   Radial shakes   Cup shakes

Effect of shakes in converted timber

Figure 4.63 Shakes

Waney edge

A waney edge is when some bark is left on the board, often because the mill has tried to maximise the amount of timber left on the log, and this remains on the cut plank. Sometimes waney edge is used as a decorative finish, such as in cladding on the outside of buildings.

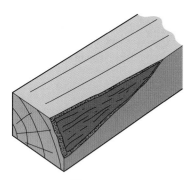

Figure 4.64 Waney edge

Bluing

Bluing, or sap staining, is a discoloration on the surface of woods such as pine and sycamore. It is a mould that has grown because the wood was stored in a moist or poorly ventilated location.

Springing

Springing is where the edge of the board is curved but the face of the board is flat. This can be caused by either poor cutting or a naturally curved grain.

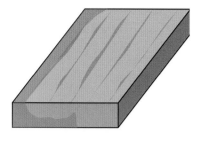

Figure 4.65 Bluing

Collapse

Collapse results in a strange shape, a smaller and malformed version of what the cut shape of the timber should look like. It is typified by rounded edges and indentations. The cells of the timber have flattened out because they have shrunk too much.

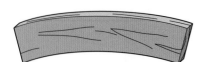

Figure 4.66 Springing

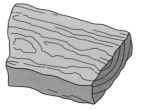

Figure 4.67 Collapse

Knots

Knots are fairly common and occur when a branch has grown out of a trunk. Knots can mean that the timber is either weakened or more difficult to work with. There are two types of knot:

* Loose (or dead) – which means that the knot is not connected to the surrounding fibres and in all likelihood it will fall out as soon as the board is seasoned.

* Live – this is where the knot is still tight and connected to surrounding fibres.

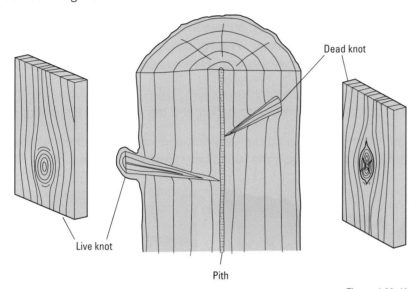

Figure 4.68 Knots

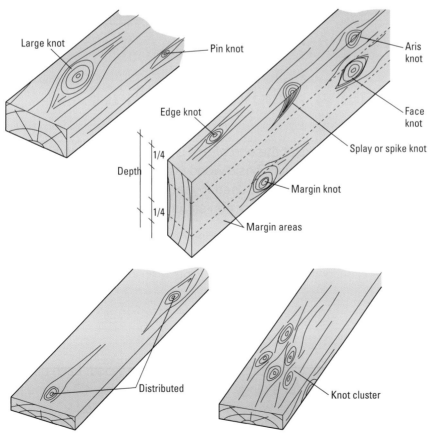

Figure 4.69 Knots and grading

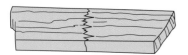

Figure 4.70 Upsets

Upsets

Upsets are when the wood fibres have been fractured across the grain. They can be caused by natural processes, such as a lightning strike or, more commonly, when the tree has been cut down and allowed to fall on something that has damaged it.

Fungal attack

Fungal attack is when microscopic strands grow and feed off the cellulose in the timber. Usually the fungus is named for the colour of the damaged timber. In hardwoods it is known as white fungus and softwoods brown fungus. But it can also be classified as being wet or dry rot. Dry rot is the most aggressive and has a musty smell. The fungus looks like cotton wool. Wet rot tends to affect exposed wood, whereas dry rot is usually seen in un-ventilated areas. The effect of these two types of rot is different:

* Dry rot makes the timber go darker and the wood cracks as it dries.

* Wet rot can leave the timber very soft and looking bleached, although some wet rot will leave the wood very flexible rather than brittle.

Figure 4.71 Dry rot

Figure 4.72 Wet rot

PRODUCING WOODWORKING JOINTS

In the practical tasks at the end of this chapter you will see how to make a variety of different basic woodworking joints. Each different type of woodworking joint is used to create common timber items that are used as either structural or non-structural parts of a building.

Over time a number of joints have been developed. Each of them has its own particular uses and all have been proven over time to work in particular situations. Mastering these joints is an important carpentry skill.

Woodworking joints and their uses

Joints increase the gluing or bonding area of the wood. Each different type of joint is used for a range of different jobs. Table 4.7 outlines the main types of joints and where they are likely to be used.

Woodworking joint	Common uses
Butt	These are simple joints that are not that strong. They can be used when joining covering materials and usually supported by either a joist or wall plate. See Fig 4.73.
Halving/Lap	These can be used to lengthen timbers and to construct flat frames. See Fig 4.74.
Angled	These can be used for a variety of different tasks, such as the decorative trims around frames of doors and windows. The joint enables two pieces of timber to be joined together to form a right angle. An example is the mitre joint, which is used to conceal the end grain and allows mouldings such as skirtings, architraves and dado rails to have a continuous unbroken finish. However, this can be overcome by reinforced dovetailed veneers or used in combination with other joints such as the mitred dovetail and the lapped mitred joint. See Fig 4.75.
Housing	This is called a stopped housing joint or a through housing joint. They can be found in staircase construction or the building of shelves. The cut may not go all the way through the timber. See Fig 4.76.
Edge	These are used to allow the use of narrow boards to either provide floor boarding or cladding. They are also used for shelving, worktops and panels. The edges of the board can either be shaped or plain. The shape of the edges of the board allows them to be interlocked, for example, tongue and grooved. See Fig 4.77.
Dovetail	These are used to create boxes and drawers. They become stronger as force is applied and the joint becomes tighter when pulled against the dovetail. The dovetail increases the gluing surface more than any other joint. See Fig 4.78.
Mortise and tenon	These are a widely used type of joint for the construction of doors and windows. There are various different shapes and types. See Fig 4.79.
Bridle	These joints are similar to mortise and tenon joints and can be used for similar purposes.
Lengthening	This is a joint that is used whenever you need to join lengths of timber to make a section long enough for the job. Types of lengthening joints include lap joints, scarfs, splayed heading and mitres. An example would be a wall plate in roofing.

Table 4.7

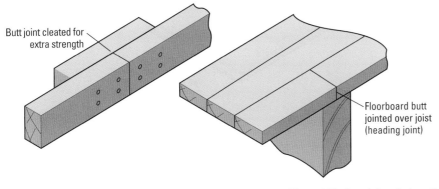

Butt joint cleated for extra strength

Floorboard butt jointed over joist (heading joint)

Figure 4.73 Butt joints in length

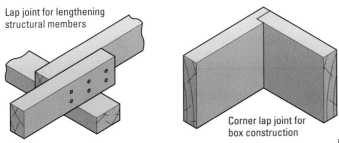

Lap joint for lengthening structural members

Corner lap joint for box construction

Figure 4.74 Lap joints

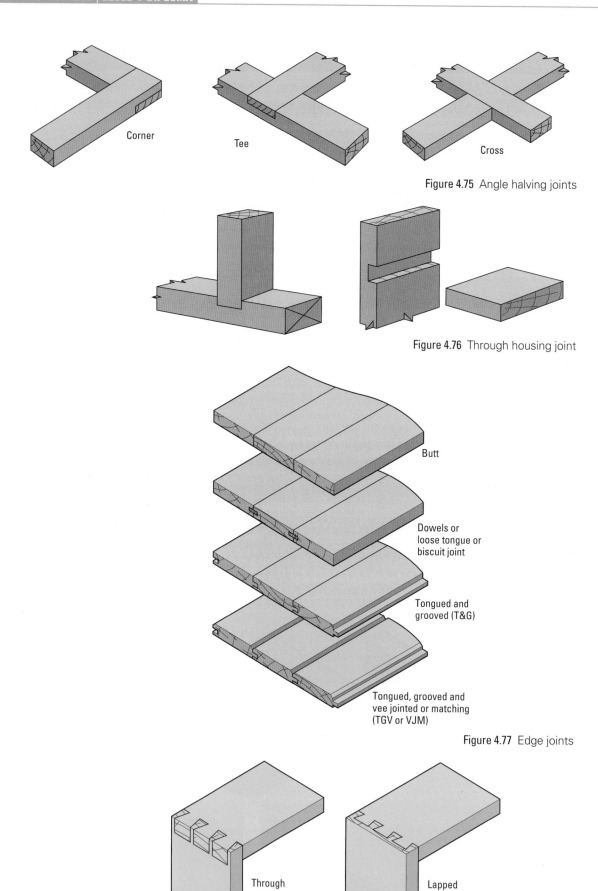

Corner

Tee

Cross

Figure 4.75 Angle halving joints

Figure 4.76 Through housing joint

Butt

Dowels or loose tongue or biscuit joint

Tongued and grooved (T&G)

Tongued, grooved and vee jointed or matching (TGV or VJM)

Figure 4.77 Edge joints

Through

Lapped

Figure 4.78 Dovetail joints

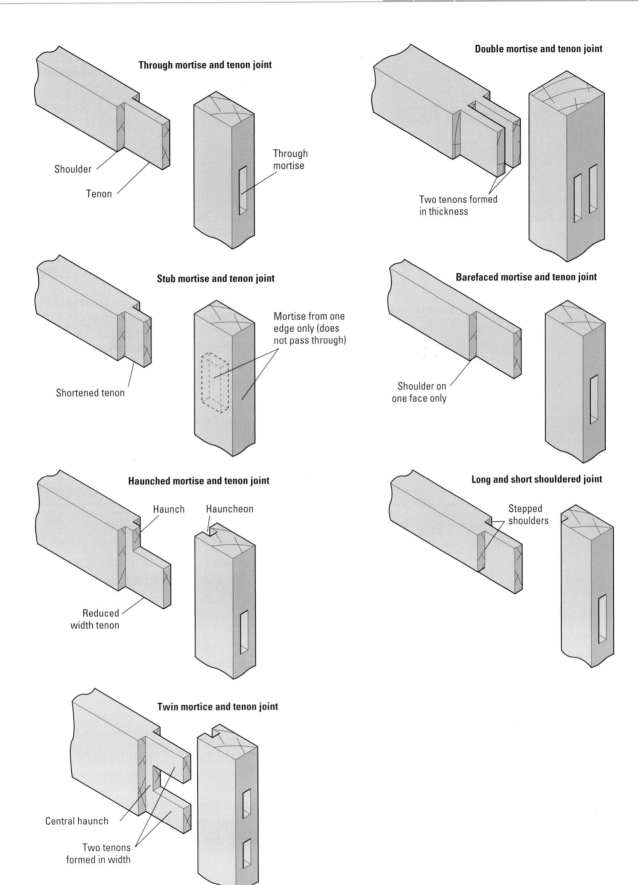

Through mortise and tenon joint

Shoulder

Tenon

Through mortise

Double mortise and tenon joint

Two tenons formed in thickness

Stub mortise and tenon joint

Shortened tenon

Mortise from one edge only (does not pass through)

Barefaced mortise and tenon joint

Shoulder on one face only

Haunched mortise and tenon joint

Haunch

Hauncheon

Reduced width tenon

Long and short shouldered joint

Stepped shoulders

Twin mortice and tenon joint

Central haunch

Two tenons formed in width

Figure 4.79 Varieties of mortise and tenon joints

Franked mortise and tenon joint

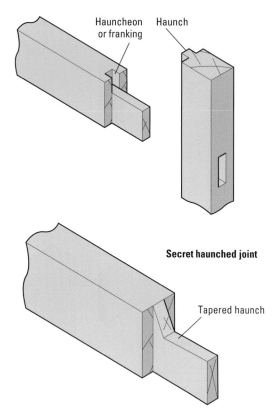

Hauncheon or franking

Haunch

Moulded frame mortise and tenon joints

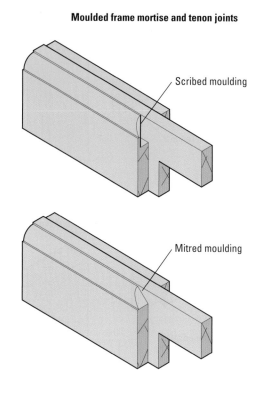

Scribed moulding

Secret haunched joint

Tapered haunch

Mitred moulding

Figure 4.79 Varieties of mortise and tenon joints *continued*

Holding and supporting

Whenever you are producing woodworking joints it makes sense to have both hands free. Also it is important to hold the timber that you are working on safely and solidly to cut down on any possible mistakes.

Exactly what you will be using to hold and support the job will depend on where you are doing it. Much of the finer work in producing joints will have to be done in a workshop. You should have access to a wide variety of holding and supporting equipment. Sometimes it is a little more difficult when you are trying to make joints on site.

Work bench

Work benches are designed with a number of features to help you hold wood in place. This will ensure that you are working on a flat surface.

The shape of the work bench is designed so that you can easily use clamps (also known as cramps). A good work bench will always have a vice, which will allow you to secure the timber.

Vices

Any good woodworker's bench should have a steel vice that is capable of opening to around 300 mm. This device holds the material against the bench in a safe and secure way. This allows you to work on the piece with both hands free. This gives you better control when using tools and makes any operation safer.

Figure 4.80 A woodworking vice

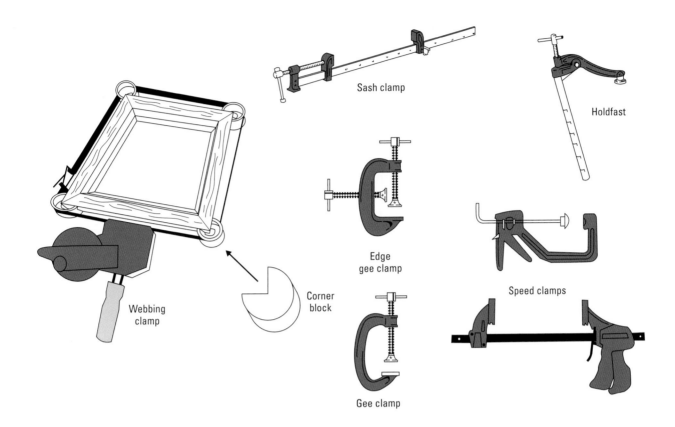

Figure 4.81 Clamps

Clamps

Clamps are designed in many different ways so that they can hold a wide variety of different materials of different sizes and shapes. Some are spring loaded or spring operated. Others have screws that tighten jaws together to secure the material. Webbing clamps have straps and a ratchet that tightens around a frame.

Frames and jigs

Jigs are templates and although they are often used to help you mark out dovetail or mitre joints they can also be used to hold material in place. Carpenters will also routinely use specialist jigs to cut timber for stair housing, hinge recesses or worktops.

Bench hook

These are used to hold small pieces of timber, particularly when cross-cutting.

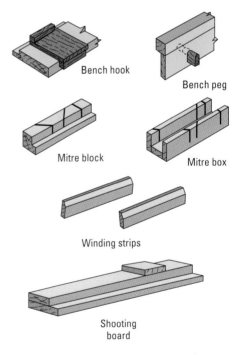

Figure 4.82 Bench equipment

Properties of adhesives

There is a wide range of different adhesives that create a chemical or mechanical bond to fix timber.

The adhesive often needs to penetrate the surface and key into the layers. This is a process known as mechanical adhesion.

The adhesive also has to have sufficient strength. This is why the adhesives tend to be applied in a liquid state. As they harden, set or cure they become solid and stronger. This is achieved in a number of different ways:

* **Solvent adhesive.** When the adhesive is applied the solvent evaporates or is absorbed into the timber.

* **Cooling.** A hot glue gun heats up the adhesive, transforming it from solid to liquid. It is applied in the liquid form but as it cools it hardens.

* **Chemical.** The adhesive needs a hardener or another chemical, or sometimes heat, to make it transform from a liquid to a solid. Synthetic resin adhesives are a good example. These are two-part powder or liquid adhesives.

* **Combination.** Some adhesives use the evaporation of solvent, a chemical reaction and cooling at the same time to make them strong.

Producing woodworking joints

The following practical tasks take you through the process of using working drawings when you are marking out. They go on to show you how to form frames using a range of different woodworking joints. The final part of this chapter gives you advice on how to dispose of waste, whether this means returning offcuts to storage for reuse or recycling the timber (perhaps reducing it to sawdust or shavings), and how to deal with general waste.

The purpose of a working drawing is to convey sufficient information to the operative to enable them to carry out the work. Setting out can be either full size (1:1) or to a suitable scale such as half full size (1:2) or quarter size (1:4). Workshop rods, whenever possible, should be marked out full size. Construction drawing practice is discussed in detail in the British Standards (BS 1192).

Before starting, make sure your work area is free of obstacles and you have all tools and materials to hand.

PRACTICAL TASK

1. MARK OUT TO PRODUCE BASIC WOODWORKING JOINTS

OBJECTIVE

To mark out from working drawings to produce basic woodworking joints.

Figure 4.83 Checking the tape measure for accuracy

Check a square by holding it against the edge of a board and marking a line. Then turn the square over and place the blade against the mark to compare.

Figure 4.84 Checking a square for accuracy

WORKSHOP SETTING OUT

It is common practice to produce a setting out rod showing the vertical and horizontal sections only. More complex products may require an elevation.

TOOLS AND EQUIPMENT

Steel rule

Tape measure

4H pencil

T-square

Set squares 45° and 30/60°

Compasses

Dividers

Trammels

Try square, combination square

Protractor

Pencil sharpener

Eraser

All drawing equipment should be checked for accuracy and square before starting to set out.

Check the tape measure against a steel rule to ensure accuracy.

SETTING OUT A HAUNCHED MORTISE AND TENON FRAME

Select a rod that is suitably large to accommodate the drawing; this can be a piece of MDF, plywood, timber or paper.

The frame in the example is 300 mm high × 230 mm wide. All components are 45 mm × 25 mm.

PRACTICAL TIP

Note that, for the purposes of clarity, the lines in the photos in this book have been made with pen instead of pencil, so are heavier than a normal construction line.

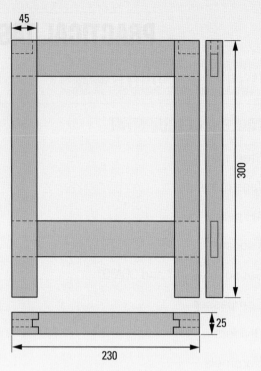

Figure 4.85 Haunched mortise and tenon frame

Ensure points are sharp on compasses and dividers.

STEP 1 Use the try square to mark a square line across the left-hand end of the rod. The line should be faint. This is called a construction line.

Figure 4.86 Marking a square line

STEP 2 Measure 300mm from the first line and square another faint line across the rod. This represents the overall length of the frame – the vertical section.

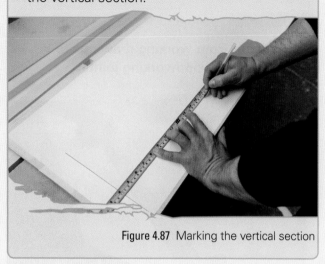

Figure 4.87 Marking the vertical section

STEP 3 Set a combination square to 25mm and, using a pencil, gauge a line down the edge of the rod. This represents the face of the vertical section of the frame.

STEP 4 Set the combination square to 50mm and gauge another parallel line down the edge of the board. This represents the back of the vertical section of the frame.

STEP 5 Measure 230mm from the first line at the left-hand end of the rod and mark another square line across the rod. This represents the overall width of the frame – the horizontal section.

Figure 4.88 Marking the back of the frame

Figure 4.89 Marking the horizontal section

STEP 6 Set the combination square to 75 mm and pencil gauge another parallel line between the width marks. This represents the horizontal face of the frame.

Figure 4.90 Marking the face of the frame

STEP 7 Set the combination square to 100 mm and pencil gauge another line between the width marks. This represents the horizontal back of the frame.

Figure 4.91 Marking the back of the frame

STEP 8 From the left-hand edge of the vertical section mark in the width of the timber section (45 mm) and square a faint line across both sections. This represents the head on the vertical section and the left-hand stile shoulder on the horizontal section.

Figure 4.92 Marking the vertical head and left-hand stile shoulder

STEP 9 From the right-hand edge of the horizontal sections, measure in 45 mm and draw a further square line. This represents the right-hand stile shoulder.

Figure 4.93 Marking the right-hand stile shoulder

STEP 10 Measure 50 mm from the right-hand edge of the vertical section and square a line across it. Then measure a further 45 mm and square another line. This represents the bottom rail.

STEP 11 Measure out the thickness of the mortise on the vertical section and, using a combination square, pencil gauge between the 45 mm line. Repeat for the tenons on the horizontal section. These are shown as broken lines denoting hidden detail.

Figure 4.94 Gauging between the lines

STEP 12 From the left-hand edge of the vertical section measure in a third of the width of the head. This represents the haunch in the mortise and is shown as hidden detail.

Figure 4.95 Marking the haunch

STEP 13 Mark the haunches onto the tenons on the horizontal section. These are equal to the thickness of the tenon and are square in section.

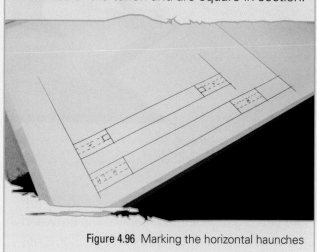

Figure 4.96 Marking the horizontal haunches

STEP 14 Mark the end grain and add a cutting list.

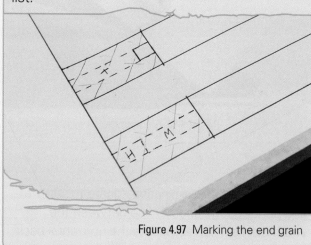

Figure 4.97 Marking the end grain

PRACTICAL TIP

Mortise and tenon joints are based upon ratios of thirds. The tenon thickness should be as close to a third of the material thickness as chisel widths allow. In this instance, an 8 mm mortise would suffice.

MARKING OUT HAUNCHED MORTISE AND TENON FRAME FROM ROD

STEP 1 Cut two stiles around 350 mm long. These should be longer than the finished length – allow approximately 30 mm for a lug at the top of the frame and 15 to 20 mm extra on the legs at the bottom.

STEP 2 Cut a head rail and bottom rail at around 240 mm to allow 5 mm of tenon to protrude through the stiles at either side.

Figure 4.98 The prepared pieces

STEP 3 Mark a face side and edge on all four pieces.

Figure 4.99 Marking face sides and face edges

STEP 4 Place one of the stiles against the vertical section on the rod, ensuring that the timber overhangs the section at both ends. Mark the positions of the mortises and the haunch and the top and bottom of the frame.

Figure 4.100 Marking the mortises

STEP 5 Place one of the rails against the horizontal section, ensuring that the timber overhangs the section at both ends. Mark the shoulder lines.

Figure 4.101 Marking the shoulder lines

STEP 6 Put the two stiles back to back with the face side pointing towards you on one stile, and the other face side facing away from you. Transfer the mortise marks and the frame length on to the second stile.

Figure 4.102 Transferring the marks

STEP 7 Square the mortise marks around to the opposite edge to the face edge. Mark the length marks on all sides and mark the waste.

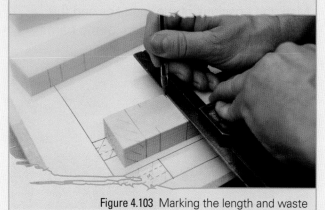

Figure 4.103 Marking the length and waste

STEP 8 Put the two rails back to back with the face side pointing towards you on one rail. Square the shoulder lines onto the second rail.

STEP 9 Square the shoulder lines around both rails on all four sides.

Figure 4.104 Squaring the shoulder lines on every side

STEP 10 Set a mortise gauge to the chisel.

Figure 4.105 Setting the mortise gauge

STEP 11 With the stock of the mortise gauge on the face side, mark the mortise holes and haunches on the stiles.

Figure 4.106 Marking the mortises

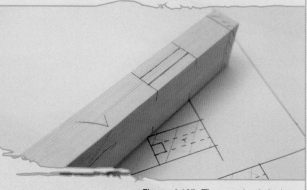

Figure 4.107 The marked timber

STEP 12 With the stock of the mortise gauge on the face side, mark the tenons on the rails.

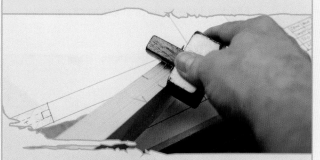

Figure 4.108 Marking the tenons on the rails

Figure 4.109 Marking the tenons across the end grain

STEP 13 Mark the depth of the haunches on the ends of the stiles using the mortise chisel as a guide.

Figure 4.110 Marking the haunches on the cheeks

PRACTICAL TIP

After the mortises have been chopped and the cheeks and shoulders of the tenons have been cut, the haunches can be marked out onto the cheeks of the tenons.

Practical tasks for forming mortise and tenon joints can be found in Chapter 5.

STEP 14 Assemble the frame.

PRACTICAL TASK

2. PRODUCE BASIC WOODWORKING JOINTS

TOOLS AND EQUIPMENT

The tools required to form the following joints can be put into the following categories:

* Measuring and marking out tools
* Shaping tools
* Parting tools
* Boring tools
* Fixing tools
* Finishing tools

Measuring and marking

* Tape measure
* Rule
* Try square
* Combination square
* Mitre square
* Marking gauge
* Mortise gauge
* Cutting gauge
* Sliding bevel

Shaping tools

* Mortise chisels
* Paring chisels
* Firmer chisels
* Internal and external gouges (inner and outer)
* Framing chisels (tools used in the construction of timber framing)
* Carpenters slicks (tools used in the construction of timber framing)

Parting tools

* Tenon saw
* Dovetail saw
* Panel saw
* Gent's saw
* Coping saw

Fixing tools

* Claw hammer
* Warrington pattern hammer
* Pin hammer
* Screwdrivers
* Nail punches

Finishing tools

* Smoothing plane
* Jack plane
* Block plane
* Shoulder plane

Holding devices

* Work bench and vice
* Bench holdfast
* G-clamps
* Bench hook
* Mitre box
* Mitre board
* Carpenter's screw clamp
* Saw horses

PPE

When producing the joints in the tasks that follow, ensure you select PPE appropriate to the job and site where you are working. Refer to the PPE section of Chapter 1.

LAP JOINTS

The lap joint is one of the simplest forms of joint; however it has many uses such as lengthening structural timbers or forming simple corners.

When lap jointing structural timbers they will be supported by a third member at the mid-point of the lap.

STEP 1 Measure and cut the timber to length.

STEP 2 Using screws or nails join the two timbers together where they overlap.

STEP 3 Fix to the supporting timber.

CORNER LAP JOINT

STEP 1 Using a rule, pencil, square and saw measure, cut the timbers to the correct length.

STEP 2 Mark the shoulder line to the thickness of the timber.

Figure 4.111 Marking the shoulder line

STEP 3 Set the marking gauge to the required depth of the lap. This could be a half or a third.

STEP 4 Using a bench hook to hold the work, cut down the shoulder line to the depth with a tenon saw.

STEP 5 Place the timber vertically in the vice and cut down to the shoulder line with a tenon saw.

Figure 4.112 Cutting down to the shoulder line

STEP 6 Assemble using glue and screws or nails.

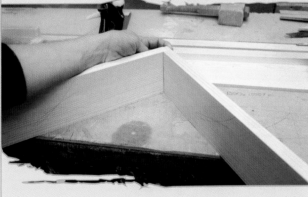

Figure 4.113 Assembling the joint

HALF LAP JOINTS

There are a number of variations of the half lap joint, including scarf joints for lengthening timbers such as wall plates in roofing, as well as corner halving, tee halving, cross halving and mitred halving. Many of these joints are used in structural framing and also in more delicate work such as furniture making.

CORNER HALVING

STEP 1 Measure and cut the timber to length.

STEP 2 Mark a face side and face edge on both pieces of timber.

PRACTICAL TIP

The stock of the square should always be on either a face side or face edge when transferring lines.

STEP 3 Measure in from one end the width of the timber and mark a shoulder line across the face side. Return this line onto both edges.

Figure 4.114 Marking the shoulder line

STEP 4 Measure in from one end of the other piece of timber and mark a shoulder line across the opposite side to the face. Return this line on to both edges.

Figure 4.115 Marking the shoulder line on the other side

STEP 5 Set a marking gauge to half the thickness of the timber. Place the stock of the gauge against the face side and mark the first piece of timber along both edges and across the end grain.

Figure 4.116 Setting the marking gauge to half the thickness

STEP 6 Place the stock of the gauge against the face side and mark along both edges and across the end grain of the second piece of timber.

STEP 7 Mark the waste on both pieces of timber.

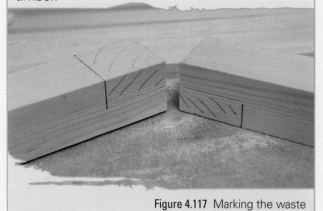

Figure 4.117 Marking the waste

STEP 8 Holding the first piece of timber in the vice, make a shallow saw cut across the end grain.

Figure 4.118 Making a shallow cut across the end grain

STEP 9 Angle the piece in the vice and cut down the end of the joint and one cheek simultaneously. Turn the piece around in the vice and repeat.

Figure 4.119 Cutting down the cheeks

Figure 4.120 Turning the piece and cutting again

STEP 10 Turn the piece in the vice so that it is vertical and complete the cut down to the depth of the shoulder line.

Figure 4.121 Completing the cut to the shoulder line

STEP 11 Remove the piece from the vice, place it on a bench hook and cut across the shoulder line to complete the first half of the joint.

Figure 4.122 Cutting the shoulder

STEP 12 Repeat Steps 8 to 11 for the second piece.

STEP 13 Secure the joint using screws and glue or nails, depending on what it will be used for.

Figure 4.123 Securing the joint

STEP 14 In joinery work the joint requires cleaning up with a plane.

PRACTICAL TIP

When producing any halving joint that forms an angle with a planed finish, it is good practice to allow a little surplus material to project past the finished length of the joint. This provides the opportunity to dress the joint without affecting the overall dimensions.

T-HALVING AND CROSS HALVING JOINTS

T-HALVING

STEP 1 Mark a face side and face edge on both pieces.

STEP 2 Mark the position of the T-joint onto the face side of the first piece of timber. This should be two square parallel lines the width of the timber apart. These are the shoulders. Square these lines onto both edges, and gauge between them, setting the gauge as before.

STEP 3 Mark the second piece of timber as in Step 4 for corner halving joint. Cut as previously described (see Steps 8 to 11 for corner halving joint).

Figure 4.124 Marking the waste

STEP 4 Place the first piece of timber in the vice and use a tenon saw to cut down the shoulders to the depth of the gauge line.

Figure 4.125 Cutting down the shoulders to the gauge line

STEP 5 Make further cuts inside the shoulders down to the gauge line. The wider the distance between the shoulders, the more intermediate cuts are needed.

Figure 4.126 Making intermediate cuts

STEP 6 Using the widest bevel-edged chisel that will fit between the shoulders, remove the waste, working from both sides of the timber towards the middle. Start with the chisel facing slightly upwards.

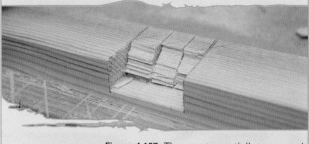

Figure 4.127 The waste partially removed

STEP 7 Pare from both sides to finish the bottom of the cut. Hold the chisel in both hands and use your index finger as a stop against the edge of the piece or the vice.

Figure 4.128 Removing the waste. Make sure you hold the chisel correctly

STEP 8 Assemble the joint as before.

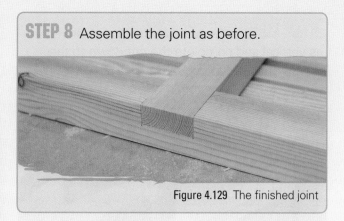

Figure 4.129 The finished joint

HOUSED OR HOUSING JOINTS
THROUGH-HOUSING JOINTS

These can be used on framing and lining construction such as doors and loft hatches. A variation of the through-housing joint is the dovetailed through-housing joint, which can be either double or single.

STEP 1 Measure and cut the timber to length.

PRACTICAL TIP

When cutting housings close to the ends of a component it is good practice to leave additional material in the length. This is known as a lug or horn. Once the housing is formed and the joint made, the lug can be removed.

STEP 2 Mark a face side and face edge on both pieces.

STEP 3 On the face side mark out the width of the housing. On a door lining this would be the thickness of the leg or jamb. Square these lines on to both edges.

Figure 4.130 Marking out the width of the housing

STEP 4 Set a gauge to the required depth and then gauge between the lines on both edges.

Figure 4.131 Gauging between the width lines

STEP 5 Using a tenon saw, cut down the shoulders.

STEP 6 Remove the waste as previously described for T-halving.

Figure 4.132 Removing the waste

STEP 7 Assemble the joint.

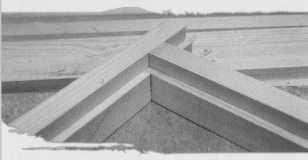

Figure 4.133 The assembled joint

BRIDLE JOINTS

The bridle is a framing joint. There are several variations, including the corner bridle, T-bridle and mitred bridle.

CORNER BRIDLE

STEP 1 Measure and cut timber to required length, allowing a little extra at the end of both pieces.

STEP 2 Mark a face side and face edge on both pieces.

STEP 3 Set a mortise gauge to the width of a mortise chisel that is closest to a third of the thickness of the material.

Figure 4.134 Setting the mortise gauge

STEP 4 Measure and mark the shoulder lines on both pieces.

PRACTICAL TIP

The male half of the joint, the tenon, will require a shoulder line on both faces and edges. The female half of the joint, the bridle, will only require shoulder lines on the edges.

STEP 5 Gauge from the shoulder lines around both edges and across the end grain of both pieces.

Figure 4.135 Gauging to mark the cheeks

STEP 6 Holding the male half of the joint in the vice, cut down the cheeks of the tenon with a tenon saw, splitting the line.

Figure 4.136 Cutting down the cheeks of the male half

STEP 7 Holding the female half of the joint in the vice, cut down the cheeks of the bridle with a tenon saw, splitting the line.

Figure 4.137 Cutting down the cheeks of the female half

STEP 8 Using a coping saw, cut to the waste side of the line and remove the centre of the bridle.

Figure 4.138 Cutting out the waste

STEP 9 Clamp the female piece down and cut back to the shoulder line using a mortise chisel cut from both sides to avoid breakout.

Figure 4.139 Chopping back to the shoulder line

STEP 10 Assemble the joint and clean up.

Figure 4.140 The assembled joint

T-BRIDLE

STEP 1 Cut timbers to length, then mark the face side and face edge on both pieces.

STEP 2 Mark out the first piece of timber as you did for the corner bridle for the female half of the joint (the bridle).

STEP 3 Measure and mark shoulder lines across the second piece of timber. This is very similar to the marking out for a T-halving; however, in this joint material is removed from the face side and the opposite face side.

Figure 4.141 Shoulder lines and gauging the marked out section

STEP 4 Gauge from the shoulder lines as before.

STEP 5 Cut the bridle as before.

STEP 6 Cut down the shoulder lines on the second piece and remove the waste as you did for the T-halving joint.

Figure 4.142 Removing the waste

STEP 7 Assemble the joint and clean up.

Figure 4.143 Assembling the joint

BUTT AND EDGE JOINTS

This type of joint is mainly used for jointing in width.

BUTT OR RUBBED JOINT

STEP 1 Arrange the boards so that the annular growth rings in the end grain alternate. This helps to minimise any cupping of the finished board.

STEP 2 Mark face side marks on the boards to be jointed and number the boards from left to right.

Figure 4.144 The boards arranged and numbered

STEP 3 Take two adjacent boards and place them in the vice, flush and back to back, face side out and corresponding edges face up.

Figure 4.145 Boards correctly positioned in the vice, side by side with face side out

STEP 4 Using a jointer, try or jack plane, shoot the edge of both boards simultaneously. Remove the boards from the vice and place them edge to edge on a flat surface to check the fit. Repeat until you achieve the fit you want.

Figure 4.146 Planing the edge simultaneously

STEP 5 Place witness marks on the face of the boards so that they can be aligned when you glue them.

STEP 6 Repeat Steps 3, 4 and 5 until all boards are complete. Dry assemble in sash clamps and check the fit.

Figure 4.147 Dry assembly

STEP 7 Place glue on the edge of each board and rub the two together to remove any air and to even out the glue. Then bring the witness marks together and place the boards in the clamps.

Figure 4.148 Rubbing the joint

STEP 8 Use a hammer and a block of timber to flush up any misaligned boards.

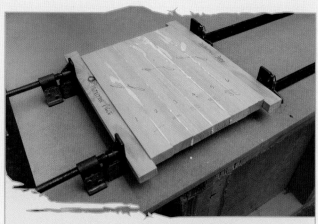

Figure 4.149 Clamped and flushed up boards

STEP 9 After the glue has set, remove the boards from the clamps and remove any excess glue from the surface using a scraper or old plane blade.

STEP 10 Secure the board to a flat surface. Use a jack or try plane to traverse across the boards diagonally. Work from one end to the other, and then repeat, traversing across the boards from the opposite corner. This operation is sometimes referred to as scrubbing.

Figure 4.150 Scrubbing, traversing across the grain

STEP 11 Finally, clean up with a jack or try plane, working along the grain.

Figure 4.151 Cleaning up, with the grain

Figure 4.152 The finished joint

LOOSE TONGUE EDGE JOINTS

The addition of a loose tongue makes this joint stronger by increasing the glue line and also aligning the boards in thickness. This reduces the amount of cleaning up required.

STEP 1 Follow Steps 1 to 6 for butt or rubbed joints (pages 139 to 140).

STEP 2 Using a plough plane, form a groove on the joint edge of each board. The depth of the groove should be slightly deeper than the loose tongue.

Figure 4.153 Forming a groove

STEP 3 Apply glue to the joint, insert the loose tongue and glue the next board. Repeat until all boards are in place.

Figure 4.154 Inserting the loose tongue

Figure 4.155 The finished joint

STEP 4 Clamp the boards as before and clean up.

TONGUE AND GROOVED EDGE JOINTS

STEP 1 Follow Steps 1 to 6 for butt or rubbed joints and then Step 2 from *Loose tongue edge joints* above.

STEP 2 Set a rebate plane to a depth that is slightly less than the depth of the groove and thick enough to be a push fit into the joint. Test it on a sacrificial piece. When a good fit is achieved run the groove by forming the two rebates on either face of the board to form the tongue.

STEP 3 Assemble and clean up as before.

MORTISE AND TENON JOINT

The mortise and tenon joint is possibly the most versatile joint used in construction. There are many variations; however, the method of setting out is very similar for all. Whenever possible, the mortise and tenon joint will be set out around a ratio of a third.

THROUGH MORTISE AND TENON: DROPPING THE MORTISE

STEP 1 Cut the components to an approximate length. When forming mortise and tenon joints it is good practice to allow extra length on both horizontal and vertical members. This allows for a horn to be left on the piece and for the tenon ends to protrude through the frame.

PRACTICAL TIP

Horns are often removed on site and provide protection to the corners of frames and doors while they are transported. Allowing tenons to stick through the frame provides a means of dressing off the tenon for a good finish and provides support to the wedges as they are driven in while clamping up.

STEP 2 Mark a face side and edge on all pieces.

STEP 3 Mark the position of the mortise on one of the pieces. In practice this could be a stile or a rail.

Figure 4.156 Marking the position of the mortise

STEP 4 Mark the shoulder lines of the tenon on the other piece.

Figure 4.157 Marking the shoulder lines

STEP 5 Set a mortise gauge to a mortise chisel. The chisel should be as close to a third of the thickness of the material as possible.

STEP 6 Gauge between the mortise lines on both edges of the piece, ensuring that the stock of the gauge always remains on the face side.

Figure 4.158 Gauging between the mortise lines

STEP 7 Gauge from the tenon shoulders, along the face edge, on to the end grain and on to the opposite edge, ensuring the stock of the gauge remains on the face side at all times.

STEP 8 Clamp the piece of timber with the mortise marked on it to the surface of the bench.

Figure 4.161 Chopping the mortise

Figure 4.159 Clamping the timber

STEP 10 Turn the piece over and chop from the opposite edge, working in the same way until the mortise is through the timber. Unclamp the piece and push the debris through the mortise. Be careful not to allow the chisel to pass all the way through the piece.

STEP 9 Use a mortise chisel to chop the mortise. Start in the middle and work out in both directions, in a V-shape, turning the chisel as the work proceeds. Do not lever off the edge of the mortise as this can damage the work – instead, push or pull the chisel towards the centre of the mortise. When the mortise is approximately half-way, turn the flat of the chisel to touch the line and square up the ends of the mortise.

Figure 4.162 Completing the mortise

STEP 11 Re-clamp the piece and square up the ends of the mortise as before.

Figure 4.160 Starting the mortise

Figure 4.163 The finished mortise

CUTTING THE TENON

STEP 1 Place the piece in the vice at an angle and cut down the cheeks of the tenon. Try to split the line (cutting to the waste side of the line will result in an overly tight tenon).

STEP 2 Turn the piece over in the vice and cut down the cheeks of the tenon from the opposite edge.

Figure 4.164 Cutting down the cheeks

STEP 3 Turn the piece so that the shoulder line is horizontal to the top of the vice and cut down the cheeks of the tenon down to the shoulder line.

STEP 4 Place the piece on a bench hook and cut the shoulders, undercutting slightly to ensure a tight fit on the face of the frame. Again, try to split the line.

Figure 4.165 Cutting down to the shoulder line, slightly undercutting

STEP 5 Test the fit by bringing the two halves of the joint together.

Figure 4.166 Test fitting the joint

STEP 6 Take the joint apart and before final assembly chop wedge room on the outside edge of the mortise.

Figure 4.167 Chopping wedge room. Note the angle of the chisel

STEP 7 Cut the wedges.

Figure 4.168 Cutting the wedges

STEP 8 When assembling as part of a frame, clamp the piece using sash clamps after applying glue to all the surfaces of the joint.

STEP 9 Check for square.

STEP 10 Apply glue to the wedges and drive home, outside wedges first, until tight.

Figure 4.169 Driving the wedges home

Figure 4.170 Cutting wedges

Figure 4.171 The finished joint

HAUNCHED MORTISE AND TENON

Where a mortise and tenon joint is used on a corner, such as in door construction, it is necessary to form a haunch to the tenon. This prevents the mortise and tenon joint from being weakened after the removal of the lugs. It would otherwise become a corner bridle joint and would be susceptible to drop on the outside edge.

STEP 1 The marking out is much the same as the previous mortise and tenon joint except the mortise will be reduced by a third of its width by the introduction of a haunch. This should be marked on the setting out. The mortise is marked onto the piece in the normal way.

Figure 4.172 Haunch a third of the mortise

Figure 4.173 The haunch is equal to the chisel thickness on the end elevation

PRACTICAL TIP

When gauging a haunched mortise and tenon, the gauge lines should extend to the ends of the lug on the face edge and onto the end grain to the depth of the haunch. The depth is equal to the width of the mortise, e.g. if the mortise is 12 mm, the haunch is 12 mm deep.

STEP 2 Chop or drop the mortise as described for a through mortise and tenon.

STEP 3 Place the piece in the vice with the haunch end of the mortise facing towards you. Use a tenon saw to cut down the gauge lines to the depth of the haunch.

Figure 4.174 The haunch end should face you

STEP 4 Remove the waste using a chisel.

STEP 5 Cut the tenon as described for a through mortise and tenon above.

STEP 6 The haunch on the tenon can now be marked directly from the mortise to the tenon or alternatively from the setting out.

Figure 4.175 Marking the tenon from the drawing or mortise

Figure 4.176 Depth of haunch

STEP 7 Cut the haunch. The wedges for the joint can be cut from the waste at this point.

Figure 4.177 Cutting the haunch

STEP 8 Test fit the joint. Chop wedge room on the mortise. Assemble as before.

Figure 4.178 The finished joint

TWIN HAUNCHED MORTISE AND TENON

This joint is used on wide rails such as mid and bottom rails in door construction. It is used to increase the gluing surface and to reduce the effect of cupping in the rail, which could result in twist or wind in the door. On a mid-rail this joint would have twin tenons with a single haunch between; on a bottom rail the joint would have twin tenons and twin haunches.

STEP 1 Mark out the mortise as before. Mark out the divisions of haunch and mortise. These will be divided equally across the width of the joint: thirds for a mid-rail and quarters for a bottom rail. Chop the mortises as before.

STEP 2 Cut and chop the haunches.

STEP 3 Cut the tenon as before.

STEP 4 Mark the haunch positions from the mortise hole or the setting out. Cut haunches as before.

STEP 5 Assemble as before.

DOVETAIL JOINTS

The dovetail joint has many variations and is most commonly associated with the manufacture of furniture. However, the dovetail joint is also used extensively in structural carpentry. It is a perfect joint for lengthening the ridgeboard of roofs and is still used in structural timber framing.

Dovetails consist of tails and pins, which are set at an angle of 1:6 for softwoods and 1:7 for hardwoods, although different ratios are sometimes used. Dovetail templates are manufactured by some tool makers. Common practice is to mark out and cut the tails first and then to mark the pins; however, some carpenters prefer to do the opposite.

THROUGH DOVETAILS

STEP 1 To set a sliding bevel to the required ratio, square a line onto a piece of timber and measure out the required ratio in units of 10 mm. For a 1:6 ratio this would be 60 mm along the edge of the timber. Starting from the square line step out 10 mm and join the end of the line to this point. Alternatively, a dovetail template can be used.

Figure 4.179 Setting out a 1:6 ratio

STEP 2 Mark a face side and face edge on all pieces.

Figure 4.180 Marking the face sides and face edges

STEP 3 Use a try square to mark a shoulder line around both pieces. The gauge should be set slightly wider than the thickness of the material.

Figure 4.181 Marking the shoulder lines

STEP 4 Place the piece of timber that is to have the tails in the vice with the end grain facing upwards.

STEP 5 Mark out the ends of the tails along the end grain, spaced evenly. The tails do not have to be the same size as the pins but consider chisels when setting out the tails as the space between the tails has to be cut back to the shoulder line with a bevel-edged chisel.

Figure 4.182 Marking the ends of the tails

STEP 6 Having squared the lines across the end grain, mark the tails onto the face of the timber using the sliding bevel or a dovetail template.

Figure 4.183 Marking the tails onto the face of the timber

STEP 7 Use a dovetail or gents saw to cut down the tails. On larger dovetails a larger saw can be used but this should be appropriate to the type and quality of the work.

Figure 4.184 Cutting down the tails

STEP 8 There are two alternative methods of removing the waste between the tails.

Method 1: Remove the bulk of the waste with a coping saw and pare back to the shoulder line with a sharp bevel edged chisel from both faces.

Figure 4.185 Removing bulk of waste

Figure 4.186 Chiselling back to shoulder

Method 2: Use a sharp bevel edged chisel to chop into the timber a couple of millimetres deep along the shoulder line between the tails, and then chop along the grain to remove waste. Continue until about half-way through, then turn the piece over and repeat for the other face.

STEP 9 Use a dovetail or gents saw to form the outside shoulders.

STEP 10 Place the other piece of timber in the vice, end grain facing upwards at the end that has been gauged.

STEP 11 Lay the tails over the end grain and line up the shoulder line with the edge of the timber in the vice. Make sure face side and edge marks match up. The tails should overhang the outside edge of the corner fractionally. Mark the pins using a sharp pencil, a marking knife or a scratch awl by drawing around the tails.

Figure 4.187 Marking the pins (1)

Figure 4.188 Marking the pins (2)

STEP 12 Square these lines onto the face and mark the waste.

Figure 4.189 Marking the waste

STEP 13 Cut the pins and remove the waste, using the same technique as for the tails.

Figure 4.190 Removing the waste

STEP 14 Assemble the joint.

STEP 15 Clean up the joint with a suitable hand plane.

Figure 4.191 Cleaning up the joint

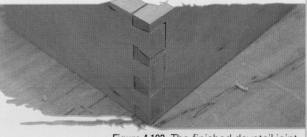

Figure 4.192 The finished dovetail joint

TEST YOURSELF

1. Which of these statements is true when gauging a haunched mortise and tenon?

   a. The width of the haunch is equal to the depth of the mortise

   b. The depth of the haunch is equal to the width of the mortise

   c. The depth of the haunch is twice the width of the mortise

   d. The width of the haunch is twice the depth of the mortise

2. Which of the following things can be done to deal with a fault or defect on a piece of equipment?

   a. Send back to manufacturer if under guarantee

   b. Send to a special repair person

   c. Worked on by yourself if the repair is minor

   d. Any of these

3. Timber can be converted in two different ways. Which of the following is the correct pair?

   a. Laterally or horizontally

   b. Tangentially or radially

   c. Horizontally or vertically

   d. Vertically or radially

4. Which type of sawing or conversion of timber aims to produce the best quality wood for joinery?

   a. Quarter sawn

   b. Tangential

   c. Through and through

   d. Rough cut

5. What should be inserted between layers of wood that you are allowing to slowly season by air circulation?

   a. Bricks

   b. Metal sheeting

   c. Sticks of the same type of wood

   d. Concrete blocks

6. Which of the following types of wood is not a hardwood?

   a. Ash

   b. Beech

   c. Redwood

   d. Teak

7. Which of the following types of timber manufactured board is made up of a series of veneers?

   a. Plywood

   b. Chipboard

   c. Hardboard

   d. Block board

8. What is a jig?

   a. A type of joint

   b. A type of vice

   c. A template that can also hold material in place

   d. A handtool used to create dovetail joints

9. Which of the following defects is discoloration or sap staining of the timber's surface?

   a. Sloping grain

   b. Bowing

   c. Waney edge

   d. Bluing

10. If timber has dry rot, what signs can be seen on the surface?

    a. A fungus that looks like cotton wool

    b. A fungus that is blue in colour

    c. The surface of the timber will look bleached

    d. There is no surface sign

Chapter 5

Unit CSA–L1Occ10
MAINTAIN AND USE CARPENTRY AND JOINERY HAND TOOLS

LEARNING OUTCOMES

LO1/2: Know how to and be able to maintain and store carpentry and joinery hand tools

LO3: Know how to use carpentry and joinery hand saws

LO4: Know how to use carpentry and joinery hand-held planes

LO5: Know how to use carpentry and joinery hand-held drills

LO6: Know how to use wood chisels

LO7: Be able to use carpentry and joinery hand tools

INTRODUCTION

The aims of this chapter are to:

* help you to maintain and store carpentry and joinery hand tools

* help you to use a range of carpentry and joinery hand tools.

MAINTAINING AND STORING CARPENTRY AND JOINERY HAND TOOLS

As you become more expert as a carpenter or joiner you will need to work with a wide variety of hand tools. Good quality tools should be looked after, so regular maintenance and sharpening is necessary. You will need to become skilled in using tools that are capable of cutting, planing, shaping, creating holes, marking off and clamping or holding timber in place while you are working on it.

Hazards, health and safety and risk assessment

Probably the greatest hazards from hand tools are as a result of them being misused or poorly maintained. Whenever a tool cannot be properly repaired it should be replaced.

There are many different ways of controlling hazards:

* Pick the tool that has the right weight, size and type of handle and grip that you need for the job.

* Always use the right tool – it should be light and comfortable to use.

* When possible, use a vice or a clamp to hold the timber in place while you work on it.

* Always try to stand or be in a comfortable position while you are working. This means being balanced with your weight evenly distributed and not over-stretching.

See the beginning of Chapter 4 for more information on using tools safely.

Sharpening hand tools

Any hand tool with a blade will need to be sharpened from time to time. This is all part of the care and maintenance of the tool.

Sharpening saws

Although for some operations hard point saws are used that cannot be
sharpened, more traditional saws need to be topped, shaped, set and sharpened.
Table 5.1 outlines the process of sharpening and maintaining a saw.

Maintenance activity	Description
Topping	This is the first stage. The teeth are likely to be at different heights as a result of wear. A file is wedged into a block of timber. The file is then drawn carefully along the length of the saw. This will make the topped edge parallel with the bottom of the saw teeth. The points of the teeth will be re-established at the shaping stage.
Shaping	First clamp the blade in a saw stock and file from the handle end to get the teeth back to their original size and shape. A file is used horizontally across at right angles to the blade. There are two types of saw teeth: rip and cross cut. Rip saw teeth should be filed at 90° so they are shaped like an inverted right-angled triangle that slopes on the back edge and is flat on its cutting edge. Cross-cut teeth are shaped like an equilateral triangle sloping front and back. When shaping the teeth, start at the handle end and shape every other tooth at 60°. Then turn the saw around and shape every other tooth that was omitted on the first pass. This should leave all the teeth the same size and shape.
Setting	This describes the width of the saw blade from the outside of the teeth on one side to the outside of the teeth on the opposite side. Saw set pliers are placed over the blade and then squeezed to set each alternate tooth. The saw is then turned around to deal with the remaining teeth. The saw set can be set to the required number of teeth per inch depending on how fine the saw is. The greater the number of teeth per inch the finer the saw and the cut it produces. Rip saws are usually set at 5 to 7 per inch and cross cuts at anything from 7 to 10 per inch. Panel saws tenon saws, dovetail saws and gents saws are set much finer.
Sharpening	A triangular saw file is used to create a sharp edge on the front of each tooth. This is an important process that will take some time to perfect. Once you have worked along the length of the saw you need to turn it round and repeat the process. Work from the handle end. The saw file will be held at 90° to the saw side for rip saw teeth and at an angle for cross-cut teeth. Look at the angle of the original sharpening of the teeth and follow the same line. Maintain the same angle on each and count the number of strokes made with the file for each tooth. Maintain the same pressure and work on every other tooth. Then turn the saw around and repeat on the other side, picking up the teeth missed on the first pass.
Dressing	This removes any final imperfections in the saw. A medium grade oil stone or slip stone is used to take off the wire edge formed by filing the teeth.

Table 5.1

Shape and size of saw teeth

Type of saw	Shape and number and types of teeth
Rip saw	The saw can be up to 750 mm long. It has 5 to 7 teeth per 25 mm. Usually this is described as being 5 to 7 tpi, or teeth per inch.
Cross-cut saw	These can be up to 650 mm long and have a tpi of 6 to 8.
Panel saw	These are around 550 mm in length and have a tpi of 10 to 12.
Tenon saw	The blades are between 250 and 350 mm long, with a tpi of 10 to 20. Many carpenters have two. They use a bevelled teeth version for cross-cutting, and one with the teeth recut square across the face for ripping with the grain.
Dovetail saw	These are between 200 and 250 mm long with a tpi of 16 to 20.
Gentlemen's saw	The blades are between 100 and 250 mm with a tpi of 32.

Table 5.2

The first diagram shows you the equipment you need to sharpen saws. The second diagram shows you how the saw maintenance is carried out.

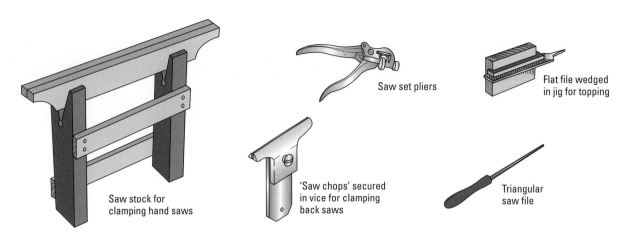

Saw set pliers

Flat file wedged in jig for topping

Saw stock for clamping hand saws

'Saw chops' secured in vice for clamping back saws

Triangular saw file

Figure 5.1 Saw sharpening equipment

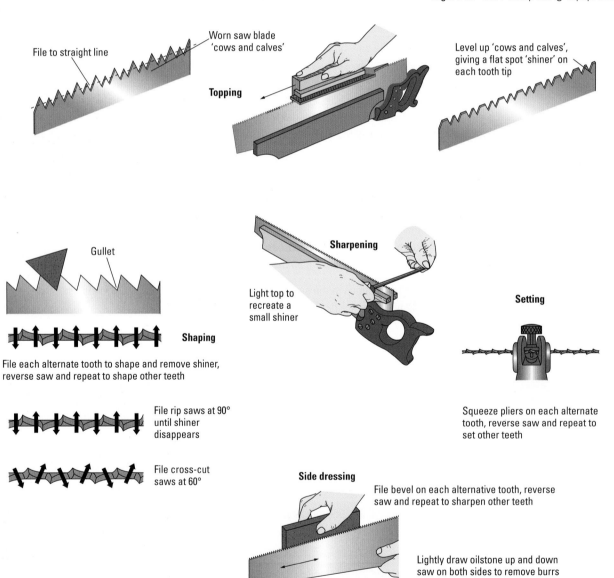

File to straight line

Worn saw blade 'cows and calves'

Topping

Level up 'cows and calves', giving a flat spot 'shiner' on each tooth tip

Gullet

Sharpening

Light top to recreate a small shiner

Setting

Shaping

File each alternate tooth to shape and remove shiner, reverse saw and repeat to shape other teeth

File rip saws at 90° until shiner disappears

File cross-cut saws at 60°

Squeeze pliers on each alternate tooth, reverse saw and repeat to set other teeth

Side dressing

File bevel on each alternative tooth, reverse saw and repeat to sharpen other teeth

Lightly draw oilstone up and down saw on both sides to remove burrs

Figure 5.2 Saw maintenance

Sharpening planes and chisels

The cutting edges of planes and chisels are ground to a **bevel** with an angle of 25°, as can be seen in Fig 5.3.

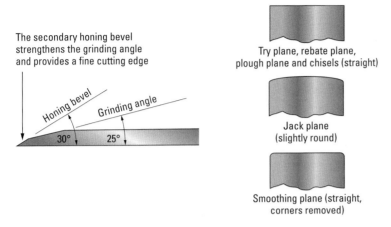

The secondary honing bevel strengthens the grinding angle and provides a fine cutting edge

Honing bevel

Grinding angle

30°   25°

Try plane, rebate plane, plough plane and chisels (straight)

Jack plane (slightly round)

Smoothing plane (straight, corners removed)

Figure 5.3 Sharpening and shaping cutting edges

The second bevel is the honing bevel and is formed at 30° to make it more durable. Some carpenters/joiners prefer to keep the grinding and honing bevels the same – this is perfectly acceptable and gives a keener or sharper edge. The only difference is that the second method will require more frequent honing of the cutting edge. The thin edge may chip more easily.

Most of the cutting edges of planes and chisels are square, although there are some exceptions. The corners of all plane blades are removed to stop them from digging in at the corners. Plane blades can be sharpened with a slight radius and some tradespeople prefer this technique, as it encourages a smoother cutting action.

When you are sharpening planes and chisels the first stage is to grind the blade using either a grindstone or the coarse side of a sharpening stone if no access to a grindstone is available. Once this has been done the blade is then honed using a fine sharpening stone.

KEY TERMS

Bevel

– this is a slope or edge that is away from either the horizontal or vertical.

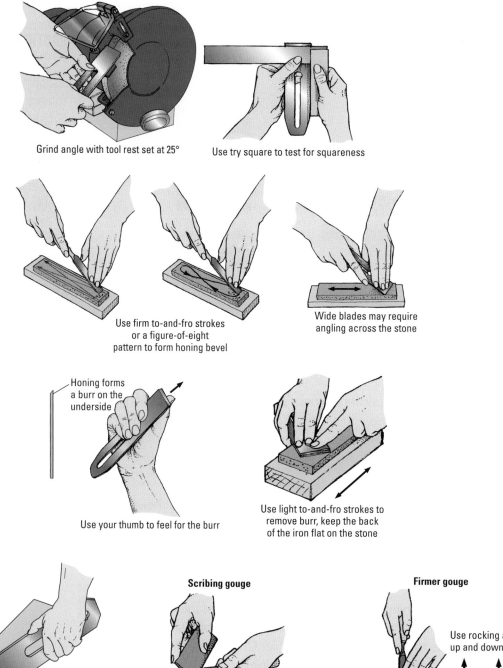

Grind angle with tool rest set at 25°

Use try square to test for squareness

Use firm to-and-fro strokes or a figure-of-eight pattern to form honing bevel

Wide blades may require angling across the stone

Honing forms a burr on the underside

Use your thumb to feel for the burr

Use light to-and-fro strokes to remove burr, keep the back of the iron flat on the stone

Draw cutting edge across corner of a piece of waste to remove wire edge left after honing

Scribing gouge

Bevel honed on slipstone

Firmer gouge

Use rocking action up and down stone

Use rocking action up and down stone

Bevel honed on flatstone

Further honing on a strop (piece of leather) will produce a super-sharp edge for working difficult grains

Burr removed on flatstone

Burr removed on slipstone

Figure 5.4 Sharpening procedure

Note that these pictures show sharpening tools being used without gloves, for clarity. You should always wear appropriate PPE when sharpening tools.

Sharpening stones

There are various different types of sharpening stones. These are described in Table 5.3

Sharpening stone	Description and use
Oil stone	These are available in fine and medium grades, although you can also buy fine and medium oil stones that are in fact two stones glued together. The medium grade is used to deal with the grinding bevel and the fine grade for the honing. They are called oil stones as oil is used to lubricate them when they are being used.
Water stone	These are finer than oil stones but available in various grades of coarseness. They need to be lubricated with water. They are primarily used for honing.
Diamond stone	Tiny particles of diamonds are bonded to a plastic base. They are available in a number of different grades. These are also used to hone an edge on the tool, providing a sharp blade.

Table 5.3

Legislation and grinding wheels

Grinding wheels present particular hazards because of the speed of their rotation and the fact that they are very abrasive. You must be sure you are fully trained and competent before using any abrasive wheel. Possible accidents include friction burns, crushed fingers and loss of eyesight.

Under the Health and Safety at Work Act (1974) employers have to make sure that risk assessments are carried out. The Abrasive Wheels Regulations were revoked and now come under the Provision and Use of Work Equipment Regulations (1998), where there are additional requirements. This means that the rules cover anyone using either fixed or portable grinding wheels and anyone who might come into contact with flying particles from the machine.

Operators of abrasive wheels should receive appropriate training and wear the right PPE. This means wearing some kind of eye or face protection, which could mean goggles or a face shield. You should also wear gloves and, to protect you against the dust, you should have respiratory protection.

However, PPE should not be the first line of defence in protecting yourself. Grinding wheels must have guards, shields and, to protect the workplace from dust and other contaminants, a local exhaust ventilation system (LEV).

KEY TERMS

Local exhaust ventilation system (LEV)

– this is a dust or fume extraction system that cleans the air.

PPE for maintaining hand tools

As an absolute minimum you should always wear eye protection when maintaining your hand tools.

When you are sharpening hand tools on a grindstone you will be creating tiny particles of metal. These are potentially very dangerous, particularly to the eyes.

You should always wear some kind of hand protection too. The thickness of the gloves will depend on the job, as you may need to be able to hold things as well. These will protect you against cuts, abrasions and any impact.

Different types of carpentry and joinery hand tools

For the purposes of this section we have listed and briefly described the different types of carpentry and joinery hand tools. Each of the hand tools is covered in more detail in the following sections of the chapter. Table 5.4 describes different types of saw.

Saws

Type of saw	Description and use
Rip saw	These are large saws, primarily used for cutting along the grain.
Cross-cut saw	These are used for cross-cutting purposes.
Panel saw	These are the shortest and finest of the hand saws and used to cut sheets.
Tenon saw	These are useful for cutting of cheeks and shoulders of tenons. It is also used for fine carpentry and joinery operations.
Dovetail saw	These are used for cutting dovetails, mouldings, detailed and fine work.
Gents saw	This is a fine back saw. With virtually no set it produces a very fine cut due to the thinness of its kerf and the number of teeth.
Bow saw	Originally this was a general site carpentry saw but more modern versions can be used for cross-cutting, ripping and cutting curves. They are used for general carcassing work and are often used in green oak timber framing.
Coping saw	Coping or fret saws are used to cut circles, curves and shapes. They were originally used in conjunction with a fretting table, but their primary use these days is to scribe mouldings such as skirtings on internal angles.
Japanese saw	These cut on the pull rather than the stroke and have long teeth and a thin blade. They are used for carcassing work as well as fine work. They can be purchased in various sizes and with varying degrees in terms of the accuracy of cut.
Keyhole saw	These are also called pad saws and are used for cutting out small shapes or holes.
Compass saw	These are used to cut out larger shapes or holes.
Mitre saw	These are used in a metal jig to cut specific angles. They are used for fine mouldings and for other small components.
Hacksaw	This is a range of different saws that are primarily used for cutting metal.
Floorboard saw	These are short saws with teeth running both in the normal way but also on a curved nose. This allows a cut to be started without first drilling through the board.

Table 5.4

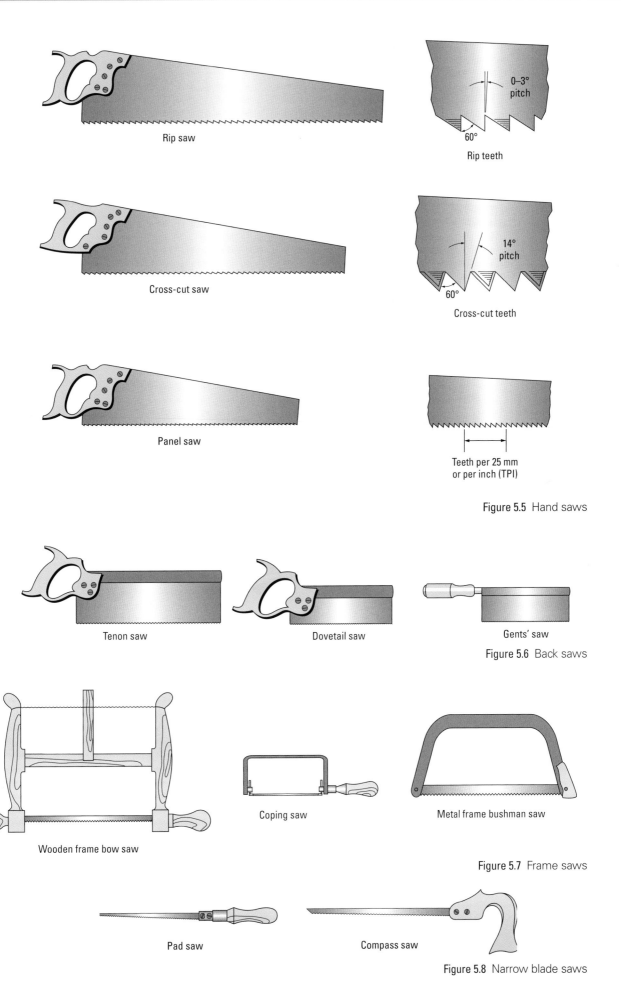

Rip saw

0–3° pitch

60°

Rip teeth

Cross-cut saw

14° pitch

60°

Cross-cut teeth

Panel saw

Teeth per 25 mm
or per inch (TPI)

Figure 5.5 Hand saws

Tenon saw

Dovetail saw

Gents' saw

Figure 5.6 Back saws

Wooden frame bow saw

Coping saw

Metal frame bushman saw

Figure 5.7 Frame saws

Pad saw

Compass saw

Figure 5.8 Narrow blade saws

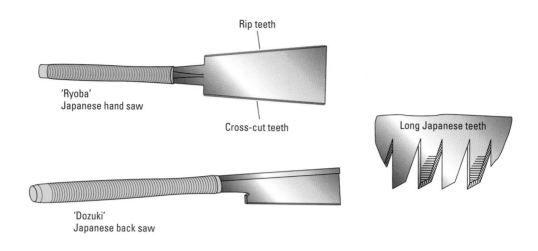

Rip teeth

'Ryoba'
Japanese hand saw

Cross-cut teeth

Long Japanese teeth

'Dozuki'
Japanese back saw

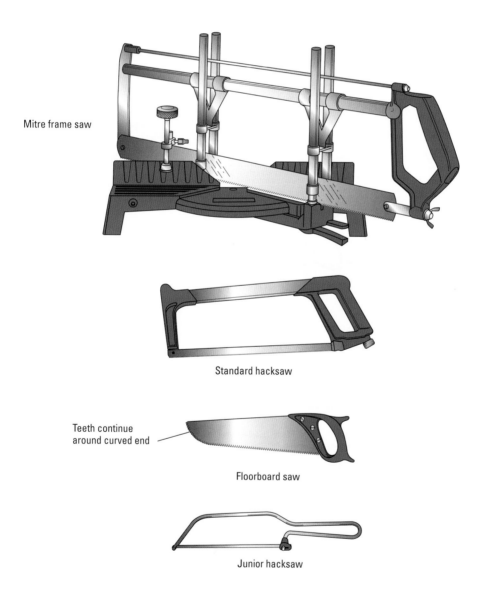

Mitre frame saw

Standard hacksaw

Teeth continue
around curved end

Floorboard saw

Junior hacksaw

Figure 5.9 Other types of saw

Chisels

At the very least, a competent tradesperson should keep in their toolbox bevel-edged chisels ranging from 6 mm to 32 mm, three or four mortise chisels, 10 mm, 12 mm, 15 mm and 25 mm, and in-cannel and out-cannel gouges of a common size such as 12 mm.

Type of chisel	Description and use
Firmer chisel	This is a general-purpose chisel with a strong blade. It has a rectangular section blade that can be driven into wood using either a mallet or the flat face of a hammer if the chisel is fitted with a shatterproof handle.
Registered chisel	These chisels have a steel band or ferrule on the end, which protects the handle when it is being hammered.
Mortise chisel	These are strong chisels with a thick, rectangular blade. It is used for chopping mortises.
Bevel-edged chisel	These are similar to firmer chisels, with a bevel on the edges of the blade. They are used for chiselling corners and are often used for dovetail joints and much finer work.
Paring chisel	These are longer and lighter versions of bevelled edged chisels. They are hand chisels so they are not hammered. Their extra length makes them easier to control when paring.
Skew-end paring chisel	These are used when you need to pare difficult corners.
Gouge	These are curved chisels. They can be used for carving, shaping and scribing. Scribing gouges are sharpened on the inside of their curve. The alternative is to grind and hone a firmer gouge on the outside of the curve for carving and for hollowing out.

Table 5.5

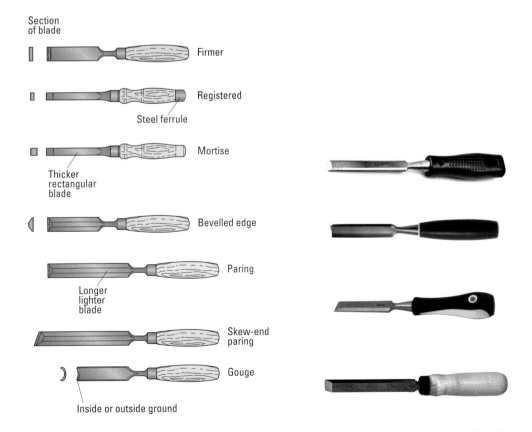

Figure 5.10 Chisels and gouges   Figure 5.11 A range of chisel handles

Planes

Type of plane	Description and use
Smoothing plane	A smoothing plane is a general purpose plane also used for cleaning up and fine finishing of timber joints.
Jack plane	A jack plane is used when greater accuracy is needed over longer lengths. The longer the plane the flatter the surface over its length. Since a jack is longer than a smoothing plane it is more accurate on longer lengths.
Try plane	Also known as a jointer plane. This is used for the truing of long lengths of timber. It is similar but easier to handle than longer planes because it is lighter.
Corrugated sole plane	These have a corrugated bottom, which is ideal if you are planing timber with a high resin content.
Bench rebate plane	Also called a carriage or badger plane, this has a blade that extends the full width of the sole. It is used for creating or cleaning large rebates.
Block plane	These are used to work across the end grain and for cleaning up laminate edges. They are sometimes used to clean up but usually only on small frames or fine work such as furniture.
Shoulder plane	Their primary function is given in the name – they are for fitting the shoulders on tenons, especially in fine cabinet work where a shoulder cut with a tenon saw does not give a sufficiently fine finish.
Bullnose plane	These are particularly useful for working into the corners where rebates meet and for stopped rebates and chamfers because their cutting iron is towards the front.
Hand router plane	These are used to level out the bottom of grooves and housings where greater accuracy is required than can be obtained with a paring chisel.
Compass plane	These are used to clean up internal and external curves and smooth them out. The sole is a flexible steel strip that can be adjusted to suit the curve.
Spokeshaves	These are used to clean up internal and external curved edges. They work best if used with the grain.
Rebate plane	These are used to cut rebates.
Plough plane	These are used to create trenches and grooves of various thicknesses and come with various interchangeable blades.
Combination or multi-plane	These are designed to be able to do the job of a range of different planes and have different cutters. This allows them to be used for ploughing, rebating and moulding.

Table 5.6

Figure 5.12 Bench plane

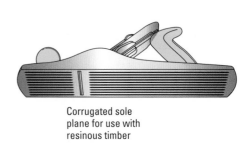

Corrugated sole
plane for use with
resinous timber

Figure 5.13 Corrugated sole bench plane

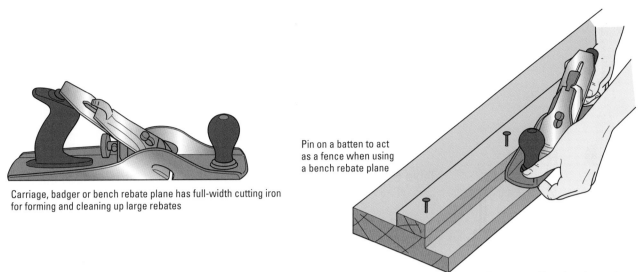

Pin on a batten to act
as a fence when using
a bench rebate plane

Carriage, badger or bench rebate plane has full-width cutting iron
for forming and cleaning up large rebates

Figure 5.14 Bench rebate plane

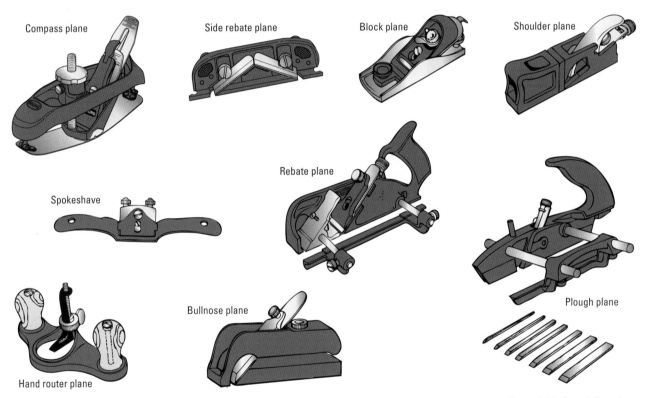

Compass plane

Side rebate plane

Block plane

Shoulder plane

Spokeshave

Rebate plane

Hand router plane

Bullnose plane

Plough plane

Figure 5.15 Specialist planes

Hand drills and braces

In the past these were the main tools used by carpenters to drill or
bore holes into timber. Even though they have been largely replaced by
battery or mains powered drills they are still very useful.

Hand drills are available in two different forms:

* **Wheel brace drills** are used for drilling and countersinking in the
  same way an electric or battery powered drill would be used.

* **Ratchet brace drills** are sometimes called carpenters braces and may have a ratchet facility. They can be used for boring out the bulk of the material when forming deep mortises for locks. By engaging the ratchet, they can be used in restrictive spaces where a full sweep of the brace is not possible. They are also used to bore holes of between 3 mm and 38 mm with standard augers and with the use of an expansive bit. This can be extended to 75 mm, which is useful for forming letter plate openings.

Table 5.7 outlines the types and use of different ratchet brace bits.

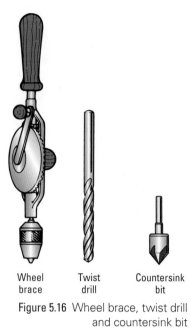

Figure 5.16 Wheel brace, twist drill and countersink bit

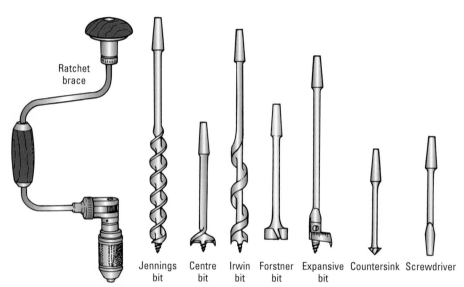

Figure 5.17 Ratchet brace and bits

Ratchet drill bit	Description and use
Jennings pattern bit	Also known as double helix bits, this feature gives a smoother finish and clears away material from the cut quicker.
Centre bit	These tend only to be used to bore very shallow holes, for example, when fitting night latches.
Irwin bit	Also known as a single helix bit, this is an alternative to the Jennings pattern bit and does the same job.
Forstner bit	These are used to create flat-bottomed holes.
Expansive bit	These are used to make large holes.
Countersink bit	These are used to create holes that will have countersunk screws in them.
Screwdriver bit	Because of the extra leverage and quicker action of the ratchet brace these are a good alternative to standard screwdrivers. They are also commonly used to extract embedded screws.

Table 5.7

Maintaining and storing hand tools

These practical tasks look at how to store and maintain hand tools. They cover all of the different resources that you will need and how to regrind angles. They also show you how to use sharpening stones, along with the PPE that you should be using.

PRACTICAL TASK

1. GRIND BLADES

OBJECTIVE

To practise grinding blades of different hand tools.

The cutting edge of plane blades, chisels and gouges can be ground to the correct angle by one of two methods: dry grinding and wet grinding. Dry grinding is a much quicker way of grinding, however, the potential for burning or blueing the blade is increased as it is much harder to keep the blade cool. When a blade blues it has effectively lost its temper at that point, resulting in a blade that will not hold an edge. Wet grinding is a much slower process but eliminates the chance of blueing the cutting edge.

There are several manufacturers of wet grinding systems using water as a coolant.

Before any work commences the machine should be correctly installed, positioned to the correct height and secured. A light mounted over the grinder is essential.

PPE

Ensure you select PPE appropriate to the job and site where you are working. Refer to the PPE section in Chapter 1.

No gloves are worn in the pictures that follow, in order to clearly show sharpening processes. However, you should wear gloves, and other PPE as appropriate, when working with sharpening tools.

WET GRINDING

STEP 1 Fill the cooling reservoir with water; this is normally a trough that the grinding wheel has to pass through. This keeps the wheel cool, stopping the build-up of heat in both the wheel and the blade. Attach the reservoir in position.

Figure 5.18 Attaching the reservoir

Figure 5.19 Attaching the fence

STEP 2 Fix the blade into the traveller. All systems incorporate a similar type of tool holder so read the manufacturer's instructions for the make and model being used.

Figure 5.20 Securing the blade in the tool holder

STEP 3 Decide if you are going to grind with the blade facing towards or away from you – both methods are acceptable. Slide the tool holder or traveller onto the bar suspended over the wheel and set the grinding angle at 25°. Most systems incorporate a method of determining and setting the correct angle.

Figure 5.21 Setting the grinding angle of 25°

STEP 4 Turn on the grinder and move the blade across the edge of the wheel. Keep a constant pressure on the blade to ensure that half the blade width is in contact with the grindstone at all times.

Figure 5.22 Beginning to grind

STEP 5 Check that the angle has been ground evenly across the blade and that the edge of the blade is square. Adjust as necessary by angling the blade in the holder.

Figure 5.23 Checking the blade for square

STEP 6 Some grinders are fitted with a leather honing wheel. It is essential that the blade is brought to this wheel with the rotation moving away from the blade to avoid a serious accident. The same holder is used as before. Apply honing paste to the wheel and hone to a sharp edge.

STEP 7 Remove the holder from the grinder and remove the blade from the holder.

STEP 8 Turn off the machine and remove the reservoir.

STEP 9 Restart the machine and allow it to run for a few minutes. This prevents the water gathering in the bottom of the wheel and making it egg-shaped.

STEP 10 The blade will have a burr on the back edge that can be felt by rubbing your thumb over the back of the blade. Apply oil to a sharpening stone and place the back edge of the plane flat on the stone, then draw the blade backwards along the stone to flatten the burr.

Figure 5.24 Honing the blade

Figure 5.25 Flattening the burr

STEP 11 Finish by stropping the blade on a piece of leather.

PRACTICAL TASK

2. HONE PLANE BLADES AND CHISELS ON AN OILSTONE

OBJECTIVE

To practise the traditional method of sharpening or honing a blade.

There are many types of sharpening stones on the market including oilstones, ceramic stones, water stones and diamond stones. The basic method of obtaining a sharp edge remains the same, the main difference being the method of lubricating the stone. Oilstone can be natural or manmade; natural stones are more expensive.

Make sure there is sufficient light and that the stone is at a good working height and secure.

PPE

In this and the tasks that follow, ensure you select PPE appropriate to the job and site where you are working. Refer to the PPE section in Chapter 1.

No gloves are worn in the pictures that follow, in order to clearly show sharpening processes.

However, you should wear gloves, and other PPE as appropriate, when working with sharpening tools.

MAINTAINING A GRINDING ANGLE ON AN OILSTONE

STEP 1 Turn the oilstone to the coarsest side. Most stones are combination stones.

STEP 2 Apply oil to the surface of the stone to prevent the stone from becoming clogged with the iron that is removed in the grinding process. Apply this as a pool at the end of the stone that is closest to you.

STEP 3 Place the blade into the pool of oil and lift the back edge until the oil is squeezed out on the front edge. At this point the grinding angle is flat to the stone.

STEP 4 Move the blade backwards and forwards down the length of the stone, keeping an even pressure. Keep your wrists parallel to the stone and avoid rocking the blade. Use the full width of the stone. Some carpenter/joiners use a figure of eight motion; however, this is not essential as long as all the stone is used.

STEP 5 Check that the blade is ground flat across the whole width of the blade by checking it for square and parallel.

STEP 6 Remove oil from the stone with a rag, and apply fresh oil.

HONING A BLADE ON AN OILSTONE

STEP 1 Apply a pool of oil to the stone and place the blade onto the stone with the back edge down. Lap the blade by moving the blade backwards, forwards and across the stone simultaneously.

STEP 2 Wipe the stone and reapply a pool of oil as before.

STEP 3 Place the blade into the pool of oil and lift the back edge of the blade until the oil squeezes from the front edge of the blade. The blade is now sitting on its 25° grinding angle. Lift the back end of the blade a little so that a 30° angle is achieved and repeat Step 4 from the previous practical task.

STEP 4 Check the cutting edge to ensure that it is parallel.

STEP 5 Check for a burr on the back edge of the blade, using the side of your thumb.

STEP 6 Place the blade flat with its back edge down on the stone and pull the blade towards you.

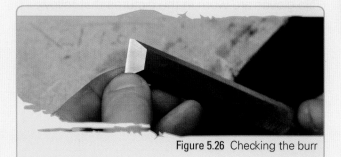

Figure 5.26 Checking the burr

STEP 7 The blade should now feel flush on the back edge and will now have a wire edge along the length of the cutting edge.

STEP 8 Strop the blade on a leather strap to remove the wire edge, leaving the edge keen.

Figure 5.27 Stropping the blade

PRACTICAL TASK

3. SHARPEN HSS DRILL BITS

OBJECTIVE

To practise sharpening drill bits.

There is a tendency to throw away drill bits that have become blunt, when in fact they should last many years with regular maintenance through sharpening. Wet grinding makes this process very simple; most systems incorporate jigs for drill sharpening to the finest tolerances. With practice a high degree of accuracy can be achieved by sharpening freehand against a tool rest.

PPE

In this and the tasks that follow, ensure you select PPE appropriate to the job and site where you are working. Refer to the PPE section in Chapter 1.

STEP 1 Check the drill manufacturer's literature to find the correct grinding angle.

STEP 2 Either set the drill sharpening jig to the correct angle using the manufacturer's instructions or use the tool rest to bring the drill bits to the edge of the stone at the correct angle.

STEP 3 When sharpening freehand, a jig to hold the drill bits should be made to keep your fingers well away from the grinding wheel.

STEP 4 Touch the drill bit against the stone at the required angle. Keep the drill bit at the same angle and twist it through 180° to grind the opposite side.

This process can be achieved through wet or dry grinding; however, wet grinding is far easier and safer.

Figure 5.28 Sharpening the drill bit

PRACTICAL TASK

4. SHARPEN AUGERS

OBJECTIVE

To sharpen augers with a file.

PPE

In this and the tasks that follow, ensure you select PPE appropriate to the job and site where you are working. Refer to the PPE section in Chapter 1.

STEP 1 Place the auger in a vice with the spurs face up.

STEP 2 The spur should always be sharpened first. Augers are manufactured to different diameters and these are measured across the outsides of the spurs. Do not file anything from the outsides of the spurs as this will render the auger useless.

Using a flat-faced file, sharpen the inside face of each spur in turn. The file should be large enough to cover the whole surface of the spur.

Figure 5.29 Sharpen the spurs first

PRACTICAL TIP

- Only file in the direction of the cut of the file.
- Do not use a forwards and backwards motion.
- Count the numbers of strokes used on the first spur and use the same amount on the opposite spur.
- Do not allow the file to touch the cutting edge of the auger.

STEP 3 Use a triangular or flat file to sharpen the cutting edge or edges (some augers have a single cutting edge but better quality augers have two). Again file in the direction of the cut of the file, keeping the file flat on the cutting angle so that it does not become rounded. Count the number of strokes used and repeat for both sides.

Figure 5.30 Sharpening the cutting edges

PRACTICAL TIP

It is important to maintain a constant even pressure when sharpening with a file to ensure a similar amount of material is removed with each stroke.

USING CARPENTRY AND JOINERY HAND SAWS

Over the years many different types of saw have been created to do specific jobs. They fall into four main categories:

* **Hand saws** are used for the first cutting to reduce the timber to the size required. Hand saws are split into three sub categories: the rip saw for cutting along the grain, the cross cut for cutting across the grain, and the panel saw, which tends to be shorter and finer toothed for cutting sheet materials or where a finer cut is required in solid timber work such as second fix.

* **Back saws** are sub-divided into tenon saws for forming tenons, general bench work and second fix carpentry. Dovetail saws are for fine joinery work. A gentlemen's saw produces a really fine cut, and is used for dovetails and delicate mouldings and beading.

* **Frame saws** are saws that generally have fairly fine blades. The blade is held in tension by a frame. They are capable of many different tasks, but uniquely they can cut curves. Frame saws include bow saws, which are used in general carpentry, particularly timber framing; coping saws, which are used extensively in second fix carpentry particularly on scribing skirtings; and fret saws, which are used to produce fretwork in conjunction with a fretting table.

* **Narrow bladed saws** are for fine work, such as making keyholes, so they are long, fine and narrow. Examples are pad saws and keyhole saws, which are used for forming keyholes and cutting small circles and for cutting openings for services such as plug sockets in plasterboards.

As we will see in this section, the saw blade needs to be designed in such a way as to prevent it from binding in the timber. Each alternate tooth is bent outwards to make the saw cut (the set). The set is wider than the rest of the saw blade. The material removed during the saw cut is called the kerf.

As you can see in Fig 5.31, the figure on the top shows that the rip saw has teeth that look like the shape of a chisel. The kerf is the saw cut and the set is the width of the teeth.

The illustration below shows a different type of saw, with cross-cutting teeth. In effect each of the teeth acts like a knife. They are designed to cut through the fibres of the timber. Two cuts are made at the same time on either side of the saw blade.

Purposes of different types of hand saw

The practical tasks at the end of this chapter will take you through different ways of cutting and shaping timber using hand saws. But it is important to understand that the way you use the saw will depend on three things:

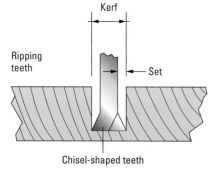

Ripping teeth

Kerf

Set

Chisel-shaped teeth

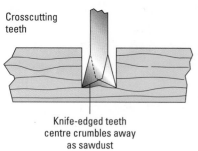

Crosscutting teeth

Knife-edged teeth centre crumbles away as sawdust

Figure 5.31 Saw teeth

* where you are doing the work – whether this is in the workshop (bench) or on site

* the direction of the cut you need to make

* the material that you are cutting.

In most cases you should follow a set procedure for using a hand saw:

1. Hold the handle with your index finger along the side. This will give you good control.

2. Put your thumbnail of the other hand on the line that you have marked to cut.

3. Begin cutting by drawing the saw backwards using short strokes.

4. Relax your grip and begin to saw using full blade strokes.

5. Apply pressure when you make a forward stroke and apply consistent pressure when pulling back.

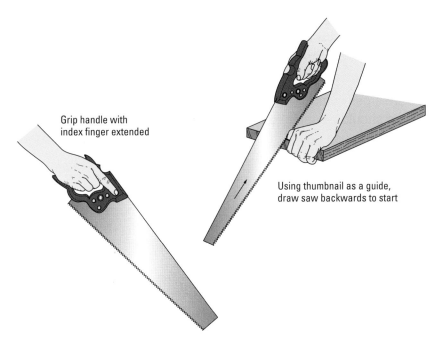

Grip handle with
index finger extended

Using thumbnail as a guide,
draw saw backwards to start

Figure 5.32 Technique for sawing

Ripping and cross-cutting on stool

For ripping on stools you should use an angle of around 60°. If you are ripping a large length or a big board you may have to balance it on two stools. If the material closes in on the blade and prevents you from using the saw in a smooth motion then you should insert a wedge. An alternative is to rub some candle wax onto the saw blade to make it run smoother.

No gloves are worn in these pictures, in order to clearly show how to use hand tools. However, you should wear gloves and other PPE required by your college or employer.

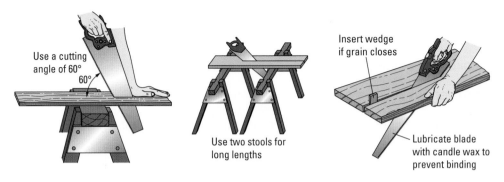

Figure 5.33 Ripping on stools

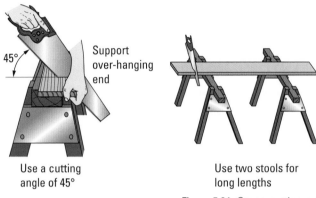

Figure 5.34 Cross-cutting on stools

If you are cross-cutting you should cut at an angle of around 45°. Any overhang should be supported. You should use your knee to secure the board and only use gentle strokes at the end of the cut, otherwise the board could split.

Ripping with the piece in a vice

Begin by putting the piece firmly into the vice and cut vertically. You will need to then undo the vice, turn the timber and finish off the cut from the other end.

If you are ripping joints it is advisable to have the timber at an angle. This will allow you to see the ripping line and the line on the end grain. You will rip down to your pencil line and this will leave the pencil line still visible.

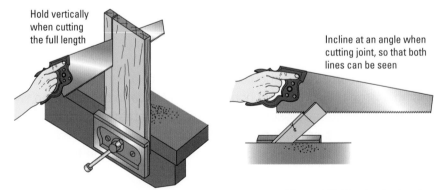

Figure 5.35 Ripping in the vice

No gloves are worn in these pictures, in order to clearly show how to use hand tools. However, you should wear gloves and other PPE required by your college or employer.

Cross-cutting on the bench

You can either hold a bench hook in a vice or hook it over the front of the bench. This serves the double purpose of protecting the bench and securing your piece of timber.

Using a coping saw

Coping saws are useful to take out waste material in joints. The piece of timber should be held securely in a vice, with around 50 mm of the timber above the surface of the bench. You can also use a coping saw to cut out shapes in timber. You begin by drilling a hole and then inserting the blade of the saw into the hole before assembling the saw.

Figure 5.36 Cross-cutting on the bench

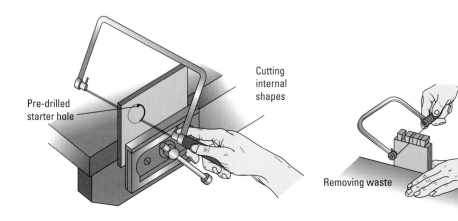

Figure 5.37 Using a coping saw

Cutting sheet material

Panel saws are useful to cut manufactured boards. If you are cutting across the grain it is often a good idea to score the faces of the board with a knife or gauge. You should make the cut on the waste side. This will prevent the board from splitting and producing a jagged edge. Boards should always be cut while fully supported, for example using a pair of saw stools. It is also advisable to put battens underneath the board for additional support.

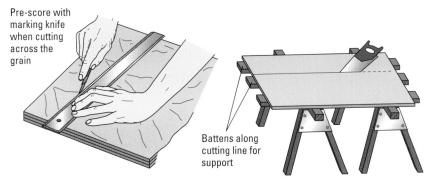

Figure 5.38 Cutting sheet material

No gloves are worn in these pictures, in order to clearly show how to use saws and hand tools. However, you should wear gloves and other PPE required by your college or employer.

CASE STUDY

LAING O'ROURKE

Using hand tools to improve your skills

Joshua Richardson is a third-year apprentice joiner at Laing O'Rourke.

'Your hand tools are definitely important. I wasn't allowed to use power tools at work until I turned 18, because they can be quite dangerous. So you have to get used to cutting everything with the hand saw. It can be a lot of work – imagine cutting down four sheets of ply at once! We still usually use hand tools for most things. One of the joiners has an electric plane, but we'll use the hand plane instead.

What my team does is maintain health and safety on the site with temporary works, so if there're holes anywhere or hoardings to go up, that's what we do.

College is totally different to my day-to-day job. It's also more about hand tools, but there it's mostly bench joinery work, learning how to make your joints, making frames, chiselling out, hanging doors, drawing, working off drawings – it's more clean, precise work. I don't get the opportunity to do much of that in my line of work, which is more site carpentry.

A lot of people at college will say, why are we doing this, why do we have to learn all this? But you might come across it one day and it gets your basic hand skills better. And, you never know, you might want to work in a different area one day where your bench joinery skills are more important.'

USING CARPENTRY AND JOINERY HAND-HELD PLANES

Planes are used to cut wood by shaving the surface. They can be used to produce flat and smooth surfaces. Others can be used for more specialist tasks. These can include making grooves, curves, rebates and mouldings.

Types and uses of planes

* **Bench planes** are used to reduce timber to its required size and shape. They can also be used to trim joints or components and to prepare the surface of timber before it is finished off.

* Specialist planes only have a cutting iron and no back iron. (The only exception is the compass plane.) The cutting iron is usually positioned bevel side up in order to break up the wood shavings, as can be seen in Fig 5.39.

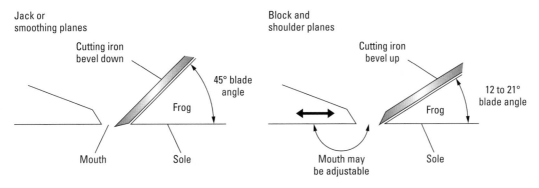

Figure 5.39 Specialist planes – cutter arrangement

The hand-held plane is arguably the most difficult tool to master. Bench planes are numbered as in the Stanley Tools system from the smallest (No. 1) to the largest (No. 8). However, they can also be described by name, for example, 'No. 4 smoothing plane', 'No. 5 jack plane', 'No. 6 fore plane', 'No. 7 try plane' and 'No. 8 jointer plane'. The most common planes are Nos 4, 4½, 5 and 6. The No. 4 is regarded as a multi-purpose plane and is used for cleaning up, although many site carpenters prefer the No. 5 or jack plane because of its suitability for fitting doors.

Planing timber

The range of different planes can be used for various jobs, including creating face sides and face edges. Each type of plane is suitable for different jobs and there are different ways of going about each task.

Using bench planes
As we can see in Fig 5.40, bench planes all have a similar type of construction. However, the thickness of the shavings that are produced and the smoothness of the finish that the plane creates are dependent on four things. These are detailed in Table 5.8.

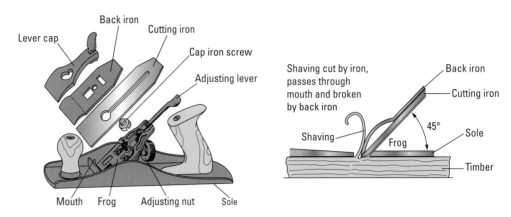

Figure 5.40 Bench plane exploded view and cutting action

Factor affecting thickness of shavings and smoothness of finish	Explanation
How far the cutting iron projects below the sole of the plane	The plane should produce clean and transparent shavings. There is an adjusting nut and you should set it so that the cutting iron projects by around 0.5 mm.
How the cutting iron is aligned	The cutting iron should be parallel with the sole. You can correct it by moving the adjusting lever sideways. The cutting iron should project an equal amount across the whole of the opening in the sole (mouth).
The distance from the back iron to the cutting iron edge	Ideally this should be 0.5 mm if you are using the plane to produce a fine and smooth finish. It can be as much as 2 mm if you are intending to take off a considerable amount of material. If there are any gaps between the back iron and the cutting iron the shavings will get jammed and the plane will clog up. You can adjust the distance by slackening the cap iron screw using a screwdriver. You can then set it to the required distance and tighten it back up again.
The size of the mouth when the cutting iron is in position	For the majority of timbers this needs to be somewhere between 1.5 mm and 3 mm. For fine work it needs to be narrow and for heavy planing it needs to be wider. The mouth size can be adjusted using the frog. You need to slacken the screws and turn the frog adjusting screw. This will move the frog either forwards or backwards. Once you have done this the securing screws need to be retightened.

Table 5.8

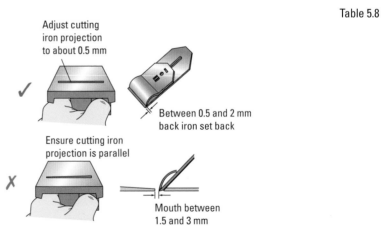

Figure 5.41 Bench plane settings

No gloves are worn in these pictures, in order to clearly show how to use hand tools. However, you should wear gloves and other PPE required by your college or employer.

You should always use jack or try planes to plane long boards. Longer planes are more accurate when planning long boards. A smoothing plane will still achieve a degree of finish and this is largely dependent on the skill of the operative. You should always try to plane with the grain. This will reduce the risk of tearing.

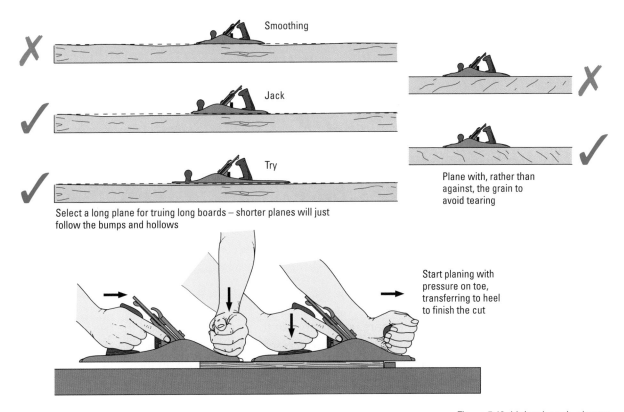

Select a long plane for truing long boards – shorter planes will just follow the bumps and hollows

Plane with, rather than against, the grain to avoid tearing

Start planing with pressure on toe, transferring to heel to finish the cut

Figure 5.42 Using bench planes

If you are working on a large area of timber you should use either a smoothing plane without corners or a jack plane, which is slightly curved. This process is shown in Fig 5.43.

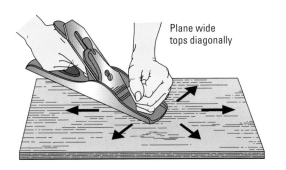

Plane wide tops diagonally

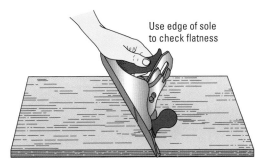

Use edge of sole to check flatness

Figure 5.43 Planing large areas

No gloves are worn in these pictures, in order to clearly show how to use hand tools. However, you should wear gloves and other PPE required by your college or employer.

Your plane should be set so that you do not remove too much material at one time. If you are uncertain as to whether or not you are taking off an even amount of material you should turn the plane onto its side. You can use the sole of the plane as a straight edge. You will then be able to see if you have any uneven parts of the surface that require more attention.

In order to plane edges you should use the centre of the sole. This is shown in Fig 5.44.

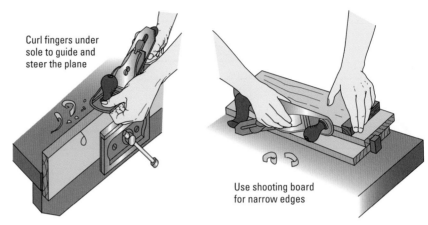

Curl fingers under sole to guide and steer the plane

Use shooting board for narrow edges

Figure 5.44 Planing edges

Use your fingers under the sole to act as a guide along the piece of timber. If you are using very narrow pieces of timber you may need to use a shooting board. This can also be seen in Fig 5.45. In this case you use the plane on its side and slide the plane along the piece of timber.

You need to choose the correct plane to clean up ready-assembled framed joinery. As you can see in the following diagram, you need to be very careful as you approach the joints. Take care that when the grain direction changes you do not tear the material. Just as you can check the flatness of a surface using the side of your plane, you can also do this for framed joinery. This will indicate whether or not you need to put any extra effort into smoothing out the area around the joints.

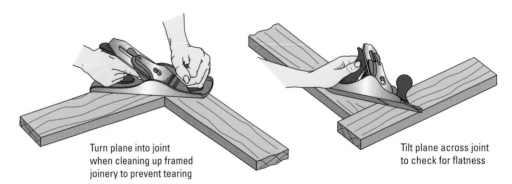

Turn plane into joint when cleaning up framed joinery to prevent tearing

Tilt plane across joint to check for flatness

Figure 5.45 Cleaning up framed joinery

No gloves are worn in these pictures, in order to clearly show how to use hand tools. However, you should wear gloves and other PPE required by your college or employer.

One of the main challenges when planing end grain is that you could cause damage in the form of breakout when you reach the end. Ideally you should plane from both corners to the centre. As Fig 5.46 also suggests you could use a shooting board.

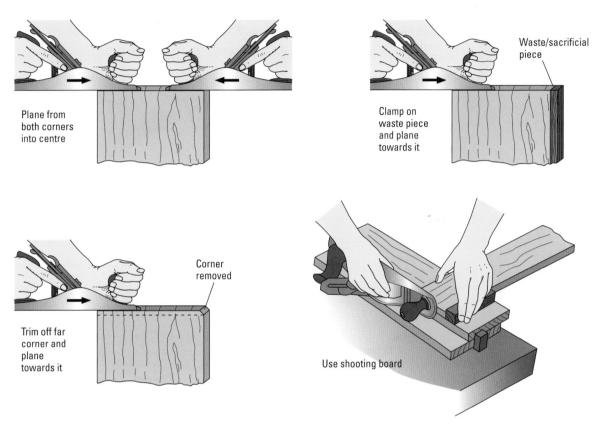

Plane from both corners into centre

Clamp on waste piece and plane towards it

Waste/sacrificial piece

Trim off far corner and plane towards it

Corner removed

Use shooting board

Figure 5.46 Planing end grain

Using specialist planes

As we have already seen there is a wide variety of different specialist planes. The ways in which these are used for a variety of different jobs can be seen in Fig 5.47 (a and b).

No gloves are worn in these pictures, in order to clearly show how to use hand tools. However, you should wear gloves and other PPE required by your college or employer.

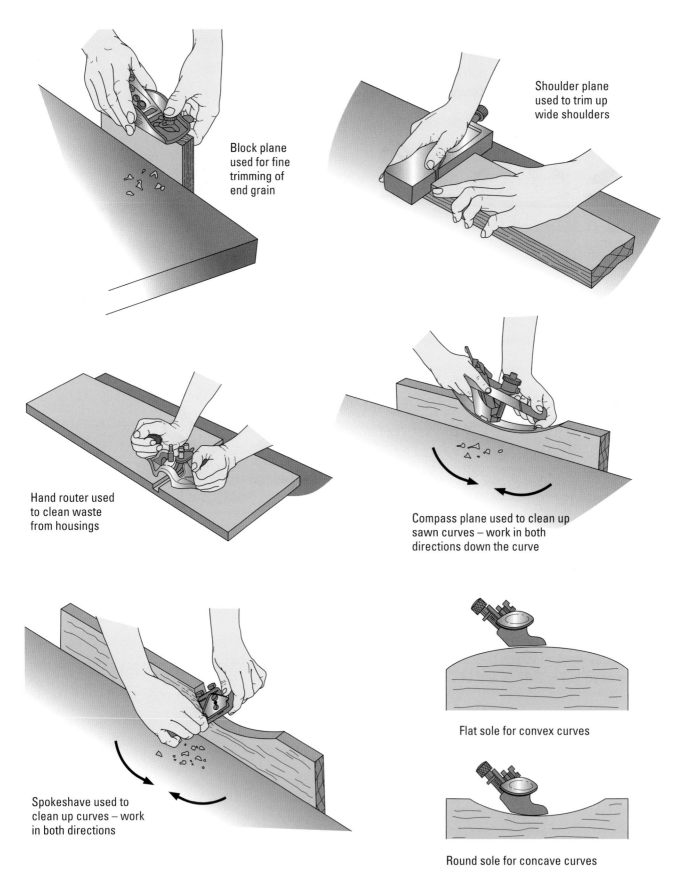

Block plane used for fine trimming of end grain

Shoulder plane used to trim up wide shoulders

Hand router used to clean waste from housings

Compass plane used to clean up sawn curves – work in both directions down the curve

Spokeshave used to clean up curves – work in both directions

Flat sole for convex curves

Round sole for concave curves

Figure 5.47a Using specialist planes

No gloves are worn in these pictures, in order to clearly show how to use hand tools. However, you should wear gloves and other PPE required by your college or employer.

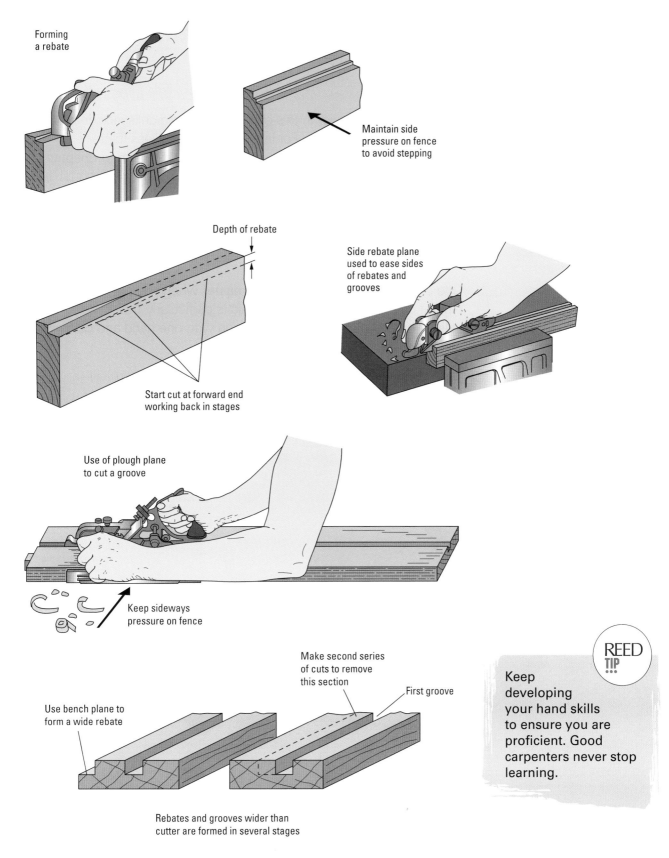

Forming a rebate

Maintain side pressure on fence to avoid stepping

Depth of rebate

Start cut at forward end working back in stages

Side rebate plane used to ease sides of rebates and grooves

Use of plough plane to cut a groove

Keep sideways pressure on fence

Make second series of cuts to remove this section

First groove

Use bench plane to form a wide rebate

Rebates and grooves wider than cutter are formed in several stages

Figure 5.47b Using specialist planes

REED
TIP

Keep developing your hand skills to ensure you are proficient. Good carpenters never stop learning.

No gloves are worn in these pictures, in order to clearly show how to use hand tools. However, you should wear gloves and other PPE required by your college or employer.

USING CARPENTRY AND JOINERY HAND-HELD DRILLS

As we have seen earlier in this chapter, hand-held drills include wheel braces and ratchet braces. For information on cordless (battery-powered) drills and modern power drills you should refer to Chapter 6, where we look at preparing and using carpentry and joinery portable power tools.

Using a range of hand-held drills

One of the most difficult things to master when you are using any type of drill is keeping it straight. This is particularly true of hand-held drills, where you are turning something with one hand and guiding with the other.

You may need another person to tell you whether you are still in alignment.

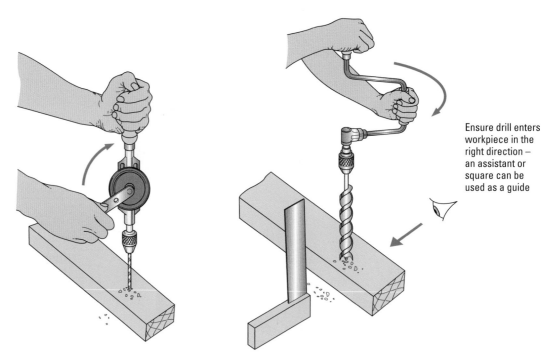

Ensure drill enters workpiece in the right direction – an assistant or square can be used as a guide

Figure 5.48 Using a drill (1)

For most carpenters a twist drill can be used for smaller holes up to around 10 mm. For larger holes they would tend to use a hand brace.

Types of bits

You should refer back to earlier in this chapter to refresh your memory about the range of bits that are used to drill holes in timber and manufactured products.

Damage to timber and the best methods of drilling through timber and manufactured products

Whenever you are drilling through timber, whether it is a board or sheet, there is always the danger of breakout or splintering on the opposite face to where you are drilling. There are two ways to avoid this:

* Drill through one face until you can see the tip of the bit appearing on the opposite side of the timber. Then reverse the timber and, using the small visible hole, position the bit on the side it came out of and complete the bore hole.

* Find a piece of wood and clamp it to the back of the piece of timber or board that you are drilling. Bore through your timber or board, straight into the waste wood underneath. If there is any splintering or damage then it will affect the waste wood and not your timber or board.

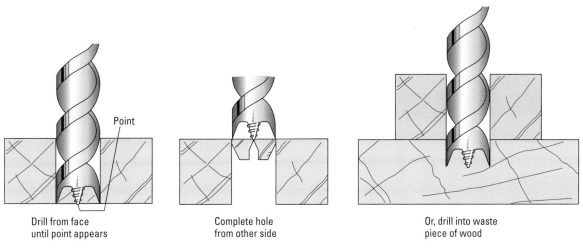

Drill from face
until point appears

Complete hole
from other side

Or, drill into waste
piece of wood

Figure 5.49 Using a drill (2)

Holding devices

Vices, clamps (clamps are also known as cramps) and jigs can be very useful in holding the material that you are working on. They give you a spare hand. Refer to Chapter 4 for more informaiton about holding devices.

USING WOOD CHISELS

As we have already seen, there are various different types of woodworking chisels. Each of these has a particular job or jobs. The other main tool that is most closely associated with a chisel is the mallet. They are made completely from wood or have rubber heads. The mallet is a joiner's tool and is used for driving chisels into timber to cut joints. Most mallets are made of beech wood.

Forming recesses and mortises using woodworking chisels

In Chapter 4 we looked at the different ways in which recesses and mortises can be created using hand-held woodworking chisels.

In the last section of this chapter, there are practical tasks that explain how to cut a mitre using a saw and then to form recesses and mortises using the appropriate wood chisel.

USING CARPENTRY AND JOINERY HAND TOOLS

This series of four practical tasks show you how to use saws, planes, drills and chisels.

It is assumed that before starting you have prepared your work area appropriately and are wearing relevant PPE including boots.

Using hand saws

Hand saws are used to cut and shape timber. This practical task will show you how to use a range of different types of saw and how to achieve different types of cuts, across and with the grain. It will also show you how to cut and shape manufactured board and how to cut mitres to given instructions.

> **REED TIP**
>
> Job interviews aren't just about finding someone who can do carpentry – hand skills are important, but it is just as important to show a positive attitude and a passion for the trade.

PRACTICAL TASK

6. USE WOODWORKING HAND SAWS

OBJECTIVE

To use a variety of woodworking hand saws.

PPE

For this and the tasks that follow, ensure you select PPE appropriate to the job and site where you are working. Refer to the PPE section in Chapter 1. No gloves are worn in these picture, in order to clearly show how to use hand tools. However, you should wear gloves and other PPE required by your college or employer.

TENON SAWS OR BACK SAWS

These are bench saws and are available in both cross-cut and rip versions. Most carpenter/joiners will only own a single tenon saw that performs both operations. They are available with between 10 and 14 tpi.

CROSS-CUTTING (TENON SAW)

STEP 1 Mark a face side and edge on the timber.

STEP 2 Measure and mark the length required and mark it with a pencil and try or combination square.

STEP 3 Place a bench hook on the bench and rest the timber against the fence.

Figure 5.50 Using a bench hook

STEP 4 Hold the tenon saw with your index finger on the side of the handle pointing towards the brass back of the saw.

Figure 5.51 Note the position of the index finger

STEP 5 Hold the timber against the fence of the bench hook with the palm of your free hand applying the pressure and your fingers holding the back of the fence. Your thumb should be free to guide the saw when starting the cut.

Figure 5.52 Holding the timber against the bench hook

STEP 6 Place your thumb against the line and the saw against the thumb; then draw the saw backwards to create a kerf in the timber on the arris closest to the fence.

STEP 7 Push the saw through the timber, applying both forward and downward pressure simultaneously. Do not try to force the saw through, but let the saw teeth do their work, using the full length of the saw.

As the cut is made, ensure your index finger on the back of the saw and the line of the cut are aligned by sighting through all three.

Figure 5.53 Sighting through the cut

PRACTICAL TIP

The saw should be sloping away from you when you start the cut. As the cut progresses the saw will flatten, and as the cut is completed the saw should be angled upwards and the speed of cut slowed.

STEP 8 When the cut is complete lay the saw down, ensuring that the teeth do not come into contact with any other tools as this will blunt the saw.

CROSS-CUTTING (TENON SAW)

STEP 1 Place the timber in the vice. For longer cuts place the timber on a saw horse and hold it in place with your knee. This should be the right knee if you are right-handed and left knee if you are left-handed.

Figure 5.54 Holding the timber with the knee

STEP 2 The timber should be held at an angle if cutting in a vice.

The saw should be held at an angle if cutting on a saw horse.

Figure 5.55 Angling the piece in the vice to make the cut

Figure 5.56 Changing the angle of the cut to avoid breakout

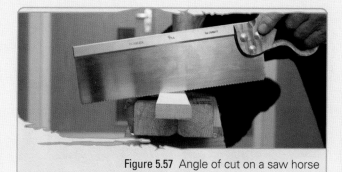

Figure 5.57 Angle of cut on a saw horse

STEP 3 With longer cuts the end of the timber should be supported as the cut progresses.

Figure 5.58 Supporting the timber

STEP 4 Follow Steps 6 to 8 for Cross-cutting, above, until the cut is complete.

RIP SAWING (HAND SAW)

STEP 1 The work should be supported across two saw horses.

STEP 2 Mark the cut on the timber.

STEP 3 Start the cut as for tenon saws, supporting the work with your knee and keeping the saw at an angle of around 60°.

Figure 5.59 Note the angle of the saw (60°)

STEP 4 When ripping long deep sections of timber it is often necessary to wedge the cut to prevent the timber trapping the blade.

Figure 5.60 Wedging the cut

STEP 5 The saw horses will need repositioning as the cut progresses.

CROSS-CUTTING (HAND SAW)

STEP 1 Place the timber across two saw horses.

STEP 2 Measure and mark the cut.

STEP 3 Hold the saw at approximately 45° and make the cut as for tenon saws. Support the weight as the cut is completed to prevent the timber from breaking and damaging the cut.

Figure 5.61 Cutting at 45°

CUTTING SHEET MATERIAL

STEP 1 Lay battens across two saw horses to provide support for the sheet.

Figure 5.62 When timber requires support

STEP 2 Lift the sheet onto the battens. For full sheets this is best done by two people to prevent injury and also to prevent potential damage to the sheet.

Figure 5.63 Marking the cut

STEP 3 Measure and mark the cut on the face side and make the cut.

Figure 5.64 The sheet on the battens to make the cut

STEP 4 The battens and saw horses will often have to be repositioned as the work proceeds.

STEP 5 Clean the cut edge using a smoothing plane and remove the splinters from the underside.

PRACTICAL TIP

It is has become common practice to use hard point saws for all the above operations. These are throw-away saws and cannot be sharpened; they are not as cost effective as a good quality, well-maintained saw that can be set and sharpened time and again throughout its life.

PRACTICAL TASK

7. USE WOODWORKING HAND-HELD PLANES

OBJECTIVE

To use woodworking planes to plane timber straight, square and to size in this and the tasks that follow.

This practical task shows you how to use woodworking planes to plane timber straight, square and to size, using a range of different types of plane.

PREPARING A FACE SIDE AND FACE EDGE

STEP 1 Examine your timber. Select the best side and edge, considering the slope of the grain and any knots.

STEP 2 Place the timber in the vice or against a bench stop with the grain running in the right direction.

Figure 5.65 Note the direction of the grain

STEP 3 Make sure the blade is parallel to the sole of the plane by aligning them using the adjustment lever. Then fully retract the blade so that there is no set on the plane.

Figure 5.66 The blade should be parallel to the sole

PRACTICAL TIP

'Set' refers to the amount of blade that extends beyond the sole of the plane. This determines the thickness of the shaving that will be removed.

STEP 4 Adopt a well-balanced stance with your weight behind the plane.

Figure 5.67 Adopting a well-balanced stance

PRACTICAL TIP

When setting up the plane in Steps 4 and 5, a scrap piece of wood can be used before starting on the actual work piece.

STEP 5 Place the plane on the work piece, applying pressure to the tote at the front of the plane and push the plane forward.

Gradually wind out the blade until it begins to bite into the work as the pass is made.

When the plane is in the centre of the piece your weight should be evenly spread at the front and back of the plane. As you get to the end of the piece transfer pressure to the handle at the back.

Figure 5.68 Transfer of weight

PRACTICAL TIP

Goggles may be required to avoid damage to your eyes from sawdust. Check with your college or employer.

STEP 6 Keep passing the plane across the piece and adjusting the blade until the required thickness of shaving is obtained. Aim to produce a continuous shaving – that is the full length of the piece and the full width of the blade. This should be as thin as possible.

Figure 5.69 A continuous shaving

STEP 7 Start by planing a face side on the piece. If the timber is rectangular in section this will be one of the widest sides.

After making sufficient passes to obtain a smooth finish the face should be checked for wind (twist).

PRACTICAL TIP

'Wind' is pronounced wined as in 'wined and dined'.

STEP 8 Place winding sticks at either end of the piece of wood and sight through the sticks. If the face is flat the sticks will be parallel. If the sticks are out of line the face is in wind.

Figure 5.70 Checking for twist

STEP 9 If the sticks are out of line, identify the high spots and remove with the plane until a flat surface that is free of wind is obtained. Finish with a single pass across the entire face.

Figure 5.71 Marking the high spots

Figure 5.72 Removing the high spots

STEP 10 Mark the face side with a face side mark.

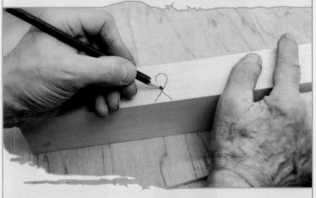

Figure 5.73 Marking the face side

STEP 11 Use the same planing technique for the face edge. The winding sticks are not used for this stage; check that the face edge is square to the face side with a try square.

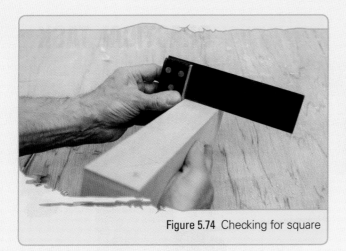

Figure 5.74 Checking for square

Figure 5.75 Marking the face edge with a face edge mark

PRACTICAL TASK

8. PLANE TIMBER TO GIVEN SIZE

Having produced a face side and face edge on the timber, it is now possible to plane the timber to a given section size.

STEP 1 Set a marking gauge to the required width with the stock of the gauge on the face edge. Mark the width on to the timber by running the gauge down the length of the timber on both the face side and the back of the timber.

Figure 5.77 Planing down to the gauge lines

STEP 3 Set a marking gauge to the required thickness. With the stock of the gauge on the face side, mark along the face edge and the back edge.

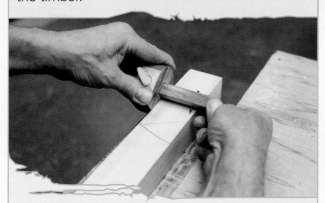

Figure 5.76 Setting off the face side

Figure 5.78 Gauge to thickness of face edge

STEP 2 Place the timber in the vice and plane down to the gauge lines until the burr begins to separate from the edge of the timber. Then make a final pass to bring the timber to size.

STEP 4 Repeat Step 2.

9. FORMING REBATES AND GROOVES

Rebates and grooves can be formed with the use of a rebate plane and a plough or combination plane respectively.

REBATES

STEP 1 Set the depth gauge and fence to the depth and width of the proposed rebate.

Figure 5.79 Setting the depth gauge and fence: rebate plane

Figure 5.80 Setting the depth gauge and fence: plough/combination plane

STEP 2 Check that the blade is projecting out from the side of the mouth of the plane by about 1 mm. This will ensure a square corner. Rebate planes are fitted with a scribing blade called a spur, which is essential when rebating across the end grain. However, it can also be used when rebating in length if required.

Figure 5.81 The blade should project by about 1 mm

STEP 3 Place the timber in the vice horizontally. Begin at the end furthest away from you and gradually work the rebate back towards you until the depth stop prevents the plane cutting further. It is essential that the plane is kept tight against the side of the work piece to prevent the rebate creeping out.

STEP 4 If the rebate is slightly out of square it can be corrected using a shoulder plane, taking care not to exceed the rebate dimensions.

Figure 5.82 Using the plane to form rebates

Figure 5.83 Forming the rebate

GROOVES

Grooves are formed in much the same way as rebates, using a plough or combination plane.

STEP 1 Set the depth stop to the required depth.

STEP 2 Set the fence to the required distance in from the edge of the piece.

STEP 3 Place the timber in the vice. Begin at the end furthest away from you and gradually work the groove back towards yourself until the depth stop prevents the plane cutting any further. Keep the fence tight against the work piece.

PLANE CURVED TIMBERS

Compass planes and spokeshaves are used to clean up internal and external curves. The compass plane is a specialised plane and most carpenters/joiners will not have one in their toolbox. Some joinery manufacturers will have a compass plane in their tool stores for general use; however they are becoming less common due to mechanisation.

STEP 1 Using the adjustment knob on the top of the plane, set the curve of the compass plane to match the internal, concaved or external curve to be cleaned.

STEP 2 Work the plane with the grain, taking care over cross grain. The blade needs to be extremely sharp to ensure a good finish. Do not attempt to remove too much material with each pass.

Figure 5.84 Using the compass plane to make a concave shape

Figure 5.85 Using the compass plane to make a convex shape

10. USE A RATCHET TO DRILL HOLES

OBJECTIVE

To use a ratchet to drill holes and produce a mortice.

STEP 1 Mark out the position of the mortise and its centre line.

STEP 2 Select an auger bit that is the width of the mortise and insert it into the jaws of the brace.

STEP 3 Mark the required depth of the mortise on to the side of the bit (auger) using masking tape or a timber dolly.

STEP 4 Secure the work piece.

STEP 5 Make sure that the ratchet is not engaged on the handle of the brace and that it is in the fixed position.

PRACTICAL TIP

The ratchet facility on a carpenter's brace is for boring holes in confined situations where the handle cannot complete a 360° sweep. The bit cuts as the handle is turned clockwise, but when this action is prevented, the handle is turned anti-clockwise and the ratchet allows the bit to stay at its correct depth. When the handle is turned clockwise again the bit cuts deeper and the process continues until the hole is bored. When the bit is to be removed the ratchet is switched to the other side and with each sweep the bit extracts itself from the hole.

STEP 6 Place the auger at one end of the mortise and bore the first hole by turning the brace handle clockwise, maintaining a constant pressure on the brace.

STEP 7 Repeat at the other end of the mortise.

STEP 8 Bore a series of overlapping holes between the outer holes.

Figure 5.86 Boring a series of holes

STEP 9 Complete the mortise with chisels.

Figure 5.87 Using chisels to finish off

Forming recesses

The following practical task discusses how to create recesses using wood chisels. It refers you to other practical tasks throughout the rest of this book that are relevant here.

PRACTICAL TASK

11. USE WOODWORKING CHISELS

The golden rule when working with any chisel is that both hands should always be behind the cutting edge. There are three main operations that are carried out with chisels:

1. Horizontal paring
2. Vertical paring
3. Chopping

Figure 5.88 Horizontal paring on a shoulder line

HORIZONTAL PARING

Horizontal paring can be executed in two ways: across the grain and with the grain.

When you are forming hinge pockets, housings, half laps and so on, the chisel should always be used across the grain.

When running in chamfers, working into rebated corners and forming gun stock shoulders the chisel is used with the grain.

For examples of using horizontal paring across the grain see the practical exercises for producing basic woodworking joints, half laps and housing joints on pages 131, 133 and 136.

VERTICAL PARING

Vertical paring is used when working across the end grain. The chisel needs to be extremely sharp.

Examples of vertical paring are the removal of waste material between tails and pins on dovetail joints and the adjustment of shoulders on tenons, although this is better done with a shoulder plane or side rebate plane.

For an example of using vertical paring across the end grain, see the practical exercise on producing basic woodworking joints for dovetails on page 147.

CHOPPING

Chopping refers to the action of cutting through timber with the chisel across the grain while working to a depth, such as in mortising. This can be part or all the way through.

Figure 5.89 Paring a mortise

The term 'chopping hinges' is often used in carpentry and joinery; however, it uses both chopping techniques and horizontal pairing.

For an example of using chopping see the practical mortise and tenons on page 142.

TEST YOURSELF

1. Which of the following is the first stage of sharpening and maintaining a saw?

   a. Setting

   b. Shaping

   c. Topping

   d. Dressing

2. How should the saw be angled when cross-cutting with a tenon saw?

   a. Keep the saw flat throughout the cut

   b. Start with the saw sloping upwards, then angle it downwards

   c. Start with the saw sloping away, flatten as you cut and end with it angled upwards

   d. Angle the saw upwards if sawing softwood and downwards if sawing hardwood

3. Which saw is useful for cutting the shoulders of joints and cross-cutting?

   a. Cross-cut saw

   b. Panel saw

   c. Rip saw

   d. Tenon saw

4. What angle should you cut at if cross-cutting on a stool?

   a. 20°

   b. 45°

   c. 60°

   d. 90°

5. Which type of plane can be used to create grooves in timber and also be used for rebates?

   a. Block plane

   b. Shoulder plane

   c. Plough plane

   d. Jack plane

6. Which type of drill bit is used with a ratchet brace to create flat-bottomed holes?

   a. Expanding bit

   b. Irwin bit

   c. Centre bit

   d. Forstner bit

7. Which of the following types of saw has the highest number of teeth per inch?

   a. Rip saw

   b. Dovetail saw

   c. Cross-cut saw

   d. Panel saw

8. What is the term used for the amount of blade that extends beyond the sole of the plane?

   a. Set

   b. Lug

   c. Waste

   d. Point

9. Which of these is not a true statement about sharpening augers?

   a. Use the same number of strokes on opposite spurs

   b. Do not use a forward and backward motion

   c. File in the direction of the cut

   d. Allow the file to touch the cutting edge of the auger

10. Which of the following is true when working with chisels?

    a. Blunt chisels are safer and easier to use

    b. Always hold the chisel in just one hand

    c. Both hands must be behind the cutting edge

    d. Chisels must never be hit with a mallet

Chapter 6

Unit CSA–L10cc11
PREPARE AND USE CARPENTRY AND JOINERY PORTABLE POWER TOOLS

LEARNING OUTCOMES

LO1/2: Know how to and be able to prepare carpentry and joinery portable power tools

LO3/4: Know how to and be able to use carpentry and joinery portable power tools to cut, shape and finish

LO5/6: Know how to and be able to use carpentry and joinery portable power tools to drill and insert fastenings

INTRODUCTION

The aims of this unit are to:

* help you to maintain and store portable power tools

* show you how to use a range of carpentry and joinery portable power tools.

PREPARING CARPENTRY AND JOINERY PORTABLE POWER TOOLS

Portable power tools can speed up many routine tasks but you must be aware of the risks of using them. According to the HSE, a quarter of all reportable electrical accidents involve portable power equipment.

Hazards and risk assessment

You should always make sure that all necessary precautions have been taken before using a power tool. Many of the points in the following list will be covered in more detail later, but it is necessary to stress their importance here:

* You must make sure that you have had training before you use any tool.

* You should make yourself familiar with the way the tool works and have practised using it in controlled conditions.

* You should give the tool a look over to make sure it is safe to use and know what to do if there is something wrong with it.

* You need to check the plug and cable for any damage.

* If you are using a mains power source, check that it's the correct voltage for the tool, and appropriate to the location of its use.

* If the power tool should have guards you should check that these are correctly fitted.

* Check any blades, drills, cutters or bits before using them. Make sure they are not worn or damaged.

* Make sure you are wearing the right PPE. This could include ear defenders, gloves, a dust mask and eye protection.

* Make sure you are not wearing any loose clothing, as it could easily become caught and injure you or make the power tool overheat.

* Never change a blade, bit, tip or drill without disconnecting the tool from the power supply first.

Risk assessment

Most risk assessments will focus on the tool itself and the electricity supply. Table 6.1 identifies the particular risks and suggests control measures.

Source of risk	Type of risk	Control measures
The power tool	1. Hair, jewellery or clothing getting tangled in moving parts 2. Eye injuries from dust or fragments 3. Hand and wrist injuries from jams and binding 4. Hand or arm vibration syndrome	1. Any loose clothing, dangling jewellery or long hair should be kept clear of moving parts 2. Suitable eye protection should be used if there is a risk of eye injury 3. Tools should only be used in accordance with the manufacturer's instructions and tested at regular intervals 4. Tools with the lowest vibration levels should be used and the amount of time individuals use the equipment should be minimised
The power supply	Electric shock	1. All power feeds should comply with European or British Standards and be in good condition 2. Where practical only 110V electrical tools should be used 3. The tools should only be used in well-lit and well-ventilated areas

Table 6.1

Usually risk assessments are graded according to the exposure to hazards. If a tool is graded as being highly unlikely to cause injury then it can be classed as a safe tool to use. However, many tools are graded as having common or even regular levels of hazard. Each of the different tools will also be graded in terms of the injury that they are likely to cause. At one end of the scale tools may produce a trivial or minor injury, perhaps a blister or a graze. At the other end of the scale they are so dangerous if something goes wrong that they could kill.

230 V mains supply

Types of power sources

There are three main types of power sources that can be used to operate portable power tools:

* either 230V or 110V electric (110V is used on construction sites)

* rechargeable battery packs

* compressed air.

Battery power pack and charger

Portable electronic generator

Portable air compressor

Figure 6.1 Power supplies

Types of portable power tools and their uses

There is a wide range of commonly used portable power tools, each of which has a different use. These can be seen in Table 6.2.

Portable power tool	Description and use
Jig saw	The blade cuts on the upward stroke on most machines, but more expensive versions have an action that moves the blade into the material on the upward stroke and away on the downward stroke. This minimises wear and tear on the blade. A number of different blades are available. They tend to be battery or mains operated at either 110V or 230V.
Drill	Drills are perhaps the most common type of portable power tool. Good quality drills can perform a variety of different actions and with the right bits, drills and other accessories they can reduce the effort needed and the time taken.
Drills (hammer and SDS rotary)	These can either be mains or battery driven. Drills are usually rated at 230V, however, if you use a transformer 110V tools can be used. On UK construction sites, all power tools have to be 110V by law. • Battery driven drills (or indeed any battery powered tools) are also known as cordless. They work using a rechargeable battery, which typically range from 12 to 36V. • There are standard rotary drills, as well as rotary or percussion drills, along with hammer drills.
Planer	A planer is used for chamfering, rebating and edging. On site it is often used to trim the edges of sheet material and for door hanging. They tend to be battery or mains operated at either 110V or 230V.
Sander	These machines aim to take much of the hard work out of finishing work. Each different type of sander leaves a more or less smooth surface, so it is important to use the right belt or paper for the job and the material being sanded. Belt sanders are good for removing stubborn defects or old paint. Orbital sanders are good for fine finished work. These tend to be battery operated at 110V.
Router	Routers have a huge number of uses – from trimming and recessing to dovetailing, drilling, moulding, rebating and grooving. They have a router cutter that can be adjusted on a spring-loaded column. They tend to be operated at 110V.
Screwdrivers	Most modern drills can also perform the functions of a screwdriver. However there are also mains and battery powered screwdrivers, using standard power supplies. Angled screwdrivers can be used where space is limited, to either remove or insert screws.
Powered nailer	These can either be pneumatic, which uses compressed air, or combustion powered, which uses a cartridge filled with a flammable gas. There are also electric powered nail guns, which either use standard mains electrical supply or are cordless and therefore battery operated.

Table 6.2

KEY TERMS

Transformer

– this is an electrical device that converts an electric current from one voltage to another. A 110V transformer should be used on construction sites.

DID YOU KNOW?

An SDS drill means special direct system. These are heavy duty drills.

Figure 6.2 Types of power sander

Figure 6.3 Types of power drill

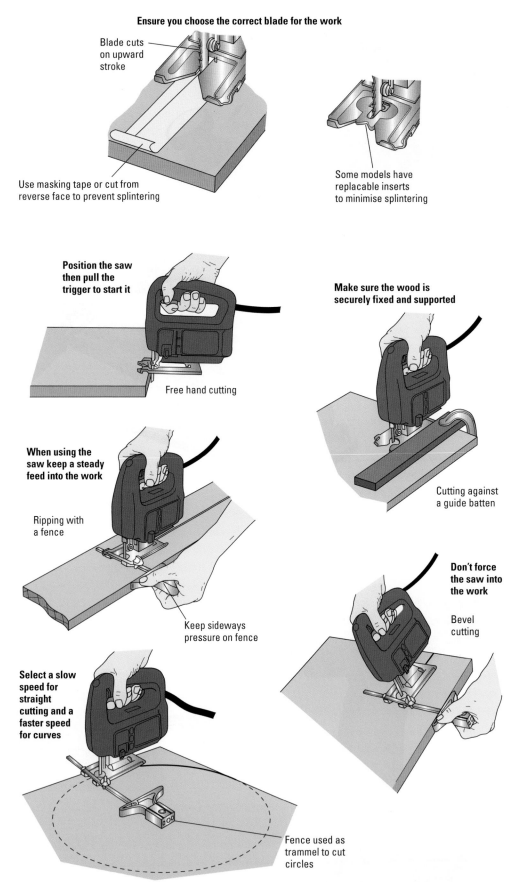

Ensure you choose the correct blade for the work

Blade cuts on upward stroke

Use masking tape or cut from reverse face to prevent splintering

Some models have replacable inserts to minimise splintering

Position the saw then pull the trigger to start it

Free hand cutting

Make sure the wood is securely fixed and supported

Cutting against a guide batten

When using the saw keep a steady feed into the work

Ripping with a fence

Keep sideways pressure on fence

Don't force the saw into the work

Bevel cutting

Select a slow speed for straight cutting and a faster speed for curves

Fence used as trammel to cut circles

Figure 6.4 Operation of a jig saw

No gloves are worn in these pictures, in order to clearly show how to use hand tools. However, you should wear gloves and other PPE required by your college or employer.

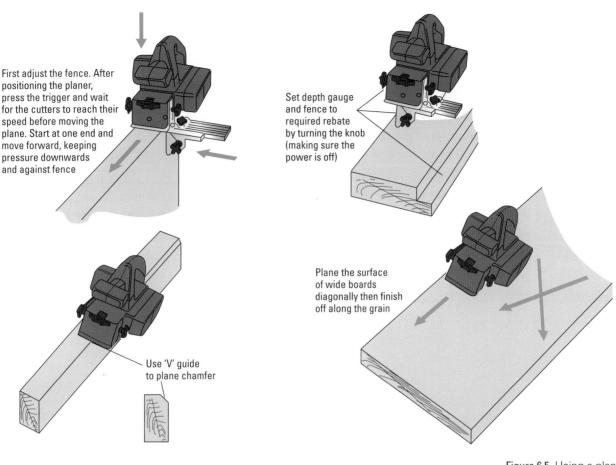

First adjust the fence. After positioning the planer, press the trigger and wait for the cutters to reach their speed before moving the plane. Start at one end and move forward, keeping pressure downwards and against fence

Set depth gauge and fence to required rebate by turning the knob (making sure the power is off)

Use 'V' guide to plane chamfer

Plane the surface of wide boards diagonally then finish off along the grain

Figure 6.5 Using a planer

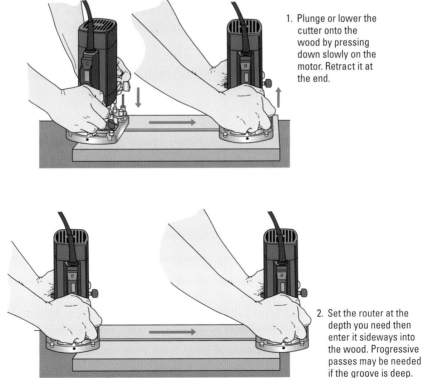

1. Plunge or lower the cutter onto the wood by pressing down slowly on the motor. Retract it at the end.

2. Set the router at the depth you need then enter it sideways into the wood. Progressive passes may be needed if the groove is deep.

Figure 6.6 Ways of using a router

No gloves are worn in these pictures, in order to clearly show how to use hand tools. However, you should wear gloves and other PPE required by your college or employer.

Drill bits

Table 6.3 shows some of the different types of drill bits that can be selected, together with their main uses.

Type of drill bit	Uses
High-speed steel (HSS)	These are drill bits that have greater resistance to heat and have greater cutting speeds. They are usually available from 2 to 13mm. They are used for drilling sheet metal or thin materials, including wood and plastic. They are said to be 4 times faster than and twice as strong as standard drill bits.
Polycrystalline diamond (PCD)	These have tiny diamond particles (about 0.5mm thick) bonded to tungsten carbide. Because diamonds are incredibly hard, they have the ability to drill through even the hardest of materials and are resistant to wear. However, they are brittle and can snap under pressure. As they are extremely expensive they are only used for particular jobs.
Disposable	Disposable bits are cheaper than standard bits because their quality tends to be poorer. However, if you are doing a lot of drilling, it may be more cost effective to spend a little more on standard reusable drill bits.

Table 6.3

PPE

Each different machine and job may require a different type of PPE. In some cases the machines produce a great deal of noise, so ear defenders may be necessary. Any machine that produces dust or particles that could fly up into your eyes, nose or mouth could require you to wear some form of eye protection. Face screens are sometimes more appropriate for better protection than a simple pair of goggles.

To protect yourself against inhaling the dust usually a dust mask is sufficient. For prolonged exposure to dust you may need to wear a respirator.

You should wear gloves where possible but some jobs, particularly when handling the machines, make it difficult to wear any kind of hand protection. If you are in doubt, speak to your trainer or supervisor straight away. If you are assisting in, for example, holding or pushing timber into a machine then you should wear gloves.

PRACTICAL TIP

Gloves are not just worn to protect against splinters and other damage. They are vital as many tools cause vibration and gloves will help prevent any long-term damage to your hands.

Maintenance and manufacturers' instructions

The majority of power tools are designed to be able to be used for long periods of time without any real maintenance. However manufacturers do recommend that power tools are checked and cleaned on a regular basis. This is because over time they could lose their efficiency.

If you are using tools that produce dust or you are working in dusty conditions it is a good idea to clean the tools daily. The dust will be attracted to the motor of the power tool and this could damage it and make it overheat.

Each manufacturer of tools provides an instruction manual. For the majority of situations this will:

* show you how to clean and maintain the power tool yourself. This will tell you what you should do on your own.

* instruct when you should get an authorised repairer to deal with the problem.

You should regularly check your power tools. This means checking for damage on the tool itself and clearing away any dust from ventilation slots. You also need to check the tool's lead and plug for any damage.

Manufacturers will also recommend how to store the power tools. In many cases this means putting them back into their case and making sure that they are not stored in wet or cold conditions.

> **PRACTICAL TIP**
>
> Don't be tempted to unscrew and take to pieces any power tools or have anyone other than an authorised repair person carry out the work. Not only are you putting yourself at risk with a poor repair, but it will also mean that any manufacturer's guarantee on the tool will be invalid.

Legislation

You should always remember health and safety legislation and take precautions when using electricity. In addition to this the following legislation also has to be considered.

Provision and Use of Work Equipment Regulations 1998 (PUWER)

These regulations aim to make sure that any machine used is safe and that it is only used for the right job. You should be trained before you use the machine and you should only ever use it if it has suitable safety measures.

The regulations cover nearly all types of equipment that you might use either in the workshop or on site. This means everything from hammers through to dumper trucks.

The most important thing to remember is that all equipment needs to be suitable for what you are using it for. It needs to be maintained and regularly inspected.

Personal Protective Equipment at Work Regulations 1992

This is a requirement that PPE is supplied and used at work if risks to health and safety cannot be controlled in another way. It covers everything from safety helmets to high visibility clothing.

The regulations state that:

* the most suitable PPE needs to be chosen on the basis that it will provide sufficient protection

* it is stored and maintained in a proper way

* you are given instructions on how to use it properly

* you actually use it in the right way.

Portable Appliance Testing (PAT)

These are visual inspections and electronic tests that check if the appliance is safe and suitable for use. It is the responsibility of the employer to make sure that appliances are inspected and tested on a regular basis. Only people that have been trained and qualified in PAT inspection can carry out these tests. Tools should not be used if they fail a PAT test.

Checking for faults and defects

As we have seen, PAT tests can be used as a regular way of checking the electrical safety of power tools. In the section on maintaining power tools we also discussed the fact that you should never use power tools that are damaged or have damaged parts. If the manufacturer's instructions say that you can replace a part then it should be fairly straightforward and safe to do this. However, you need to make sure you know what you are doing before you try it. If you are unclear you should ask someone more experienced.

If the machine is damaged in any way and you cannot repair it then you must make sure that an authorised repair person deals with the problem.

In addition to checking the tool for faults or defects before you use it, the following are examples of good practice:

Figure 6.7 Electrical PAT safety test

* There should be a regular maintenance programme and a maintenance log kept. This might mean daily, weekly or monthly checks.

* At least every 12 months a competent person should carry out a PAT test and records of these inspections and tests should be kept. The frequency of PAT tests is dependent on the tool's risk factor and how much it is used.

Preparing carpentry and joinery portable power tools

The following practical exercise looks at how to prepare power sources for power tools. It also covers how to check the tools, cables and tooling, as well as how to change tooling following the manufacturer's instructions.

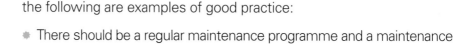

PRACTICAL TASK

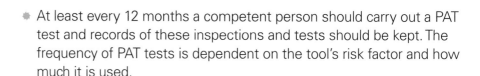

1. MAINTAIN CARPENTRY AND JOINERY POWER TOOLS

OBJECTIVE

To know how to maintain portable powered machinery.

For tasks on using and storing power tools, refer to the practical tasks from p210.

The tools used could be electrically powered, battery powered or air powered.

ROUTER CHECKLIST

* Check the power supply.
* Check the power cord is in good condition with no fraying or cracks.
* Make sure the router has been PAT tested.
* Check the overall condition of the router: is it in good condition?
* Check that all stops and guides are functioning correctly.
* Make sure that you are fully trained in the use of such machines.

MAINS RADIAL ARM SAW CHECKLIST

* Check the power supply.
* Check the power cord is in good condition with no fraying or cracks.
* Make sure the saw has been PAT tested.
* Check the overall condition of the saw: is the guarding in place?
* Check the condition of the blade and ensure it is the correct blade for the machine.
* Check that holding down cramps are in position.
* Make sure that you have been trained in the use of the particular saw.

JIG SAW CHECKLIST

* Check the power supply.
* Check the power cord is in good condition with no fraying or cracks.
* Make sure the jig saw has been PAT tested.
* Check the overall condition of the jig saw.
* Check that the correct blade is properly fitted.
* Check that the base is secure and square to the blade.
* Make sure that you are fully trained in the use of such machines.
* Check that the speed and reciprocating action are set correctly.

ORBITAL SANDER CHECKLIST

* Check the power supply.
* Check the power cord is in good condition with no fraying or cracks.
* Make sure the sander has been PAT tested, if applicable.
* Check the overall condition of the sander.
* Check that the base is in good condition with no hollows etc.
* Check that the holding down cramps for the sanding sheets are working correctly.
* Ensure the sanding sheet is securely held in the cramps and that it is flat to the base.
* Make sure that you are fully trained in the use of such machines.
* Check that the speed and reciprocating action are set correctly.

PORTABLE POWERED CIRCULAR SAW CHECKLIST

* Check the power supply.
* Check the power cord is in good condition with no fraying or cracks.
* Make sure the circular saw has been PAT tested, if applicable.
* Check the overall condition of the circular saw.
* Check the condition of the blade and that it is the correct type.
* Check that the guarding is operating correctly.
* Make sure that you are fully trained in the use of such machines.

USING CARPENTRY AND JOINERY PORTABLE POWER TOOLS TO CUT, SHAPE AND FINISH

At the end of this section a practical exercise covers how to select and use the right tooling for different jobs. It also looks at selecting the right PPE, how to use holding devices, such as vices and cramps, and then how to go about cutting, shaping and sanding different kinds of timber.

In this first part of the section we look at the fact that tooling such as blades, bits and tips can be damaged over time. We also discuss the fact that many of these machines produce debris, which in itself is hazardous.

Tooling and damage

Regardless of the cost of the tip, blade or bit the tooling will eventually have to be changed. It can wear, fracture or become rough and far less accurate.

There is not always a good way of knowing how long the tooling will last. A chip can gradually become more serious and this may depend on the roughness or the strength of the material that you are cutting, moulding, shaping or sanding.

Damaged blades, for example, will reduce the cutting speed and probably make the cut far rougher than it should be. Worn drill bits will require a higher speed and more pressure to drill holes into material.

If you buy good quality drill bits and make sure that you use the right speed and lubricate them, they can last without re-sharpening for a long time. By slowing down the speed of the drill it might take a little longer to drill the hole, but you will not have to change the bits so often. It is the speed of the drill that causes the heat and it is the heat that causes the damage to the drill bit.

Most jig saws will come with standard blades. These might work well for simple DIY projects, but at work you will need a wider range of different jig saw blades for different jobs. They are classified by how many teeth they have per inch. Always make sure that you choose the right blade for the right material. Jig saw blades are thin and are only really supported at the top, where they are attached to the jig saw. This means that they can bend and overheat quite easily. Bending is more likely to happen if you are cutting through timber that is too thick.

High-speed jig saw blades do not actually move any faster than normal blades but they do have greater penetration and can be used on thin metals. You should only ever use cobalt steel blades for thicker wood and metals. These blades can be sharpened several times before you need to replace them.

Many manufactured boards, such as plywood, include glue in their manufacture. These can reduce the life of tooling. The glue is very hard compared to the rest of the material and it dulls blades and drill bits.

If you are drilling or cutting through hardwood, such as oak, then tooling can get dull very quickly. This is also true if the wood has any knots.

Lubricants have a wide variety of different uses. There are a number of traditional types, such as oil or paraffin, but increasingly plant oil is being used. The key benefits of using lubricants are:

* they are rust inhibitors (they stop rust)

* the blade is cleaner

* the times between having to sharpen the blades are longer

* the blade tension is retained for longer.

Debris, hazards and the importance of a clean working area

Efficient dust collection is vital for health reasons but also to comply with the law. Power tool users can suffer from allergic reactions, which can affect the nose, eyes and skin. A build-up of dust can also pose a fire hazard. Larger particles that cling to surfaces can cause scoring and there is also the problem of reduced visibility, which can make accurate measurement and cuts impossible.

A simple dust collection system uses a duct system. This moves the dust from the saw to a collection device that is attached to the ducting. Metal ducting is usually thought to be better than plastic piping. This is for three reasons:

* There is a limited choice of suitable plastic pipe fittings that would meet the needs of the extraction.

* The elbows in plastic pipes tend to clog.

* Plastic piping is non-conductive – it builds up a static charge as the charge particles pass along it. This charge can shock and there is also the risk of explosion or fire.

Spiral, steel pipe with fittings that have a long radius are less likely to clog. They can also be fitted with sections that can unclip and be cleaned out. The pipe is conductive and is less likely to be a fire hazard.

Nearly every task will produce some waste, whether it is dust or small pieces of wood. The Building Act (1984) clearly states that it is the construction industry's responsibility to prevent and control waste. It should also make sure that resources are not wasted unnecessarily.

The Building Regulations cover the problem of waste disposal. This is in Part H, which covers all types of building materials, including wood.

The drive towards sustainable and secure buildings also aims to control waste and to protect the environment. You could refer back to Chapter 3, to refresh your memory about sustainability.

In order to reduce the amount of waste the Table 6.5, which follows, can be used as a guide.

KEY TERMS

Non-conductive

– this is a material that does not readily conduct electricity, and static electricity may build up in the material.

Conductive

– this means that an electrical current can pass through the material and will not build up in it.

Waste reduction and disposal method	Explanation
Elimination	This involves not producing the waste in the first place. Regularly checking stocks of materials on site or in a workshop stops over-ordering. Using cutting lists means you can order the right lengths of materials and reduce the waste. Once the materials have arrived, if they are stored properly they will be in a good state for another job. It is also important to cut as accurately as you can to minimise accidental offcuts.
Reduction	Always keep materials in their protective packaging and try not to handle the material unless necessary as this will avoid damage. Always put materials back into storage. If you have large offcuts set them aside as they might be useful later. Always use up opened stock before breaking into a new package.
Re-use	Use offcuts for pegs, profile boards and repairs. Re-use timber offcuts as many times as you can. They can be used for hoardings or form work.
Recycle	Most timber can be recycled. Some has a high value, such as reclaimed oak or pine for furniture. Most other timber, no matter how small the offcut, can be used to produce chipboard or MDF. You should throw wood into a skip only as a last resort.

Table 6.4

Cutting, shaping and sanding

This practical task looks at how to select and use the right tooling for manufacturing a two panelled door, including the fitting of a mortise lock and night latch, using the following power tools:

- Radial arm saw
- Portable powered planer
- Circular saw
- Portable router
- Drill
- 110V transformer
- Jig saw
- Orbital sander

It would not be normal practice to produce a door with these tools; however, this is an example of the versatility of this machinery and some of the ways they can be used. For this reason, no plans or measurements are given for these tasks. Each different job may need a different holding device, such as a vice, cramp or jig.

PRACTICAL TASK

2. CUT, SHAPE AND SAND TIMBER AND MANUFACTURED BOARD

OBJECTIVE

The objective for this and the tasks that follow is to use and store different power tools.

Perform your pre-use checks for the radial arm saw and circular saw, as shown in the previous task.

PPE

In this and the tasks that follow, ensure you select PPE appropriate to the job and site where you are working. For all practical tasks in this chapter, refer to the PPE section in Chapter 1. No gloves are worn in these pictures, in order to clearly show how to use hand tools. However, you should wear gloves and other PPE required by your college or employer.

STEP 1 Set out the door from the rod. Mark the position of all mortise and tenons. Remember rebated stiles and rails will require long and short shoulders. See *Setting out from rods* in Chapter 6.

STEP 2 Plug the radial arm saw into the mains supply via a 110 volt transformer.

STEP 3 Using PSE timber of stock sizes, cut to the required lengths, allow for horns on the stiles and some additional length on the rails, using a chop or radial arm saw or circular saw.

STEP 4 The saw should be cleaned down after use and a visual check should be made to ensure the saw has not sustained any damage during use. The saw and transformer should be stored in a dry, secure store with the cable properly coiled.

PRACTICAL TASK

3. FORMING THE MORTISE HOLES USING A PORTABLE POWERED ROUTER

To form the mortises, a powerful router that accepts half-inch cutters will be required.

* Perform your pre-use checks for the router.

STEP 4 Plug the router into the mains supply via a 110 volt transformer.

STEP 1 Make a simple jig that restricts the movement of the router to the outer edges of the required mortise.

STEP 5 Plunge the router to a depth of approximately 10 mm and make the first pass, continue to increase the depth with each pass until the depth is just below half way.

Figure 6.8 Making a simple jig with a jig saw

Figure 6.9 Using the router

STEP 2 Ensure the router is fitted with a cutter that has a diameter equal to the mortise width and will reach just over half the width of the stiles.

STEP 6 Turn the stiles over and repeat until the mortises are completely through the stiles.

STEP 3 Cramp the stiles securely to a bench or suitable work surface.

STEP 7 Form the slots for the haunches to the depth required. Set the depth stop on the router to make sure all haunches are equal. Always allow the cutter to stop before removing from the piece.

PRACTICAL TIP

It may be necessary to clamp additional pieces of timber of the same section either side of the piece to offer a better bearing for the base of the router.

STEP 8 Square up the ends of the mortises with a sharp chisel.

STEP 9 Clean the router after use and make a visual check to ensure the router is not damaged.

Store the router and transformer in a purpose-made box or carrying case, usually supplied with them. Always ensure that the cable is properly coiled. Place in a dry secure store.

PRACTICAL TASK

4. FORMING THE REBATES USING A PLANER

STEP 1 Set the depth stop to the depth of the required rebate. The depth stop is located on the side of the planer and is usually secured with a wing nut or similar.

STEP 2 Fit the rebating fence to the planer. This is usually connected by sliding bars that slot into the body of the planer. Set the fence to the width of the rebate.

Figure 6.10 Fitting the rebating fence

STEP 3 Set the depth of cut by adjusting the in-feed bed on the sole of the planer.

Figure 6.11 Adjusting the in-feed bed

STEP 4 Plug the planer into the mains supply via a 110 volt transformer.

STEP 5 Make several passes until the depth stop bottoms out on the surface of the stile or rail. Complete all rebating before moving on to the forming of the tenons.

Figure 6.12 Using the planer

PRACTICAL TIP

Always allow the planer to stop before removing from the piece.

STEP 6 Clean down the planer after use and give it a visual check to ensure it has not been damaged. Pay particular attention to the blades: are there any chips, are they fully tightened or is there a build of resin?

The planer and transformer should be stored in a purpose-made box or the carrying case supplied with them, with the cable properly coiled. This in turn should be kept in a dry secure store.

PRACTICAL TASK

5. FORMING THE TENONS USING A RADIAL ARM SAW

Perform your pre-use checks for the radial arm saw.

Some radial arm saws have adjustable stops that allow the saw to be pulled across the timber at a given depth. This allows the saw to be pulled across the piece to form the shoulders. By moving the piece along in small increments, tenons can be formed.

STEP 1 Mark the cheeks of the tenons with a mortise gauge.

STEP 2 Lay the timber face side down and set the depth stop on the saw.

STEP 3 Make sure the rails are clamped down using the cramps provided on the saw.

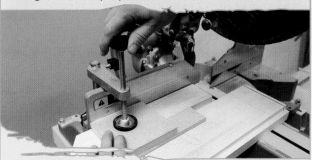

Figure 6.13 Clamping the rails

STEP 4 Cut the shoulders of the tenons on both sides. Set up a stop so all the shoulders are cut in line. Remember the shoulders will be long and short to allow for the rebate.

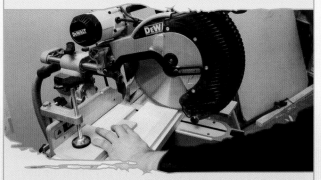

Figure 6.14 Cutting the shoulders

STEP 5 Make a series of passes over the timber to form the tenon. Clean up the cheeks with a badger or shoulder plane. Repeat for all tenons.

Figure 6.15 Forming the cheeks

Figure 6.16 Cleaning up the cheeks

PRACTICAL TIP

Alternatively tenons can be quite easily formed with the use of a router and a basic jig. Once the shoulders have been formed all that remains is for the waste to be removed with a series of passes.

STEP 6 The saw and transformer should be stored as described for the previous tasks.

6. CUTTING THE HAUNCHES USING A JIG SAW

Perform your pre-use checks for the jig saw.

STEP 1 Mark out the haunches onto the cheeks of the tenons.

STEP 2 Plug the jig saw into the mains supply via a 110 volt transformer.

STEP 3 Clamp the stiles to a suitable work surface.

STEP 4 Using a spacing piece to sit on the cheek of the tenon, cut out the haunches on each of the stiles.

Figure 6.17 Cutting the haunches

STEP 5 Before putting the jig saw away cut enough wedges from a piece of scrap timber for each tenon on the frame.

Figure 6.18 Wedges cut from scrap

PRACTICAL TIP

Always allow the saw blade to stop before removing from the piece.

STEP 6 Clean the jig saw down after use and make a visual check that it has not sustained any damage during use.

Store the jig saw and transformer in a purpose made box or carrying case, usually supplied, with the cable properly coiled. This in turn should be kept in a dry secure store.

7. ASSEMBLING THE FRAME

STEP 1 Dry fit the frame and place in sash cramps, check the fit of all joints and check for square.

Figure 6.19 Dry fitting the frame

STEP 2 Take the door out of the cramps and chop wedge room on all mortises using a sharp mortise chisel of the correct width.

Figure 6.20 Chopping wedge room

STEP 3 Apply glue to both cheeks of each tenon and along the shoulders, and assemble the door.

STEP 4 Place in the sash cramps and square as before.

STEP 5 Apply glue to each wedge and insert using a hammer.

Figure 6.21 Inserting the wedges

STEP 6 Remove from the cramps.

STEP 7 Allow glue to set.

PRACTICAL TASK

8. CLEANING UP USING AN ORBITAL SANDER

Perform your pre-use checks for the orbital sander.

PRACTICAL TIP

Some orbital sanders do not use cramps to secure the sanding sheets to the base of the machine. Instead they use self-adhering sheets. These are matched to the particular make and model and are more expensive than buying rolls of sandpaper.

STEP 1 Ensure the door is on a flat surface and held securely.

STEP 2 Fit a piece of 80 grit sandpaper into the sander.

Figure 6.22 Fitting the sandpaper

STEP 3 Plug the sander into the mains supply via a 110V transformer.

STEP 4 Start with the high points on the door surface; these are most likely to be at the joints. Work along the grain as this will reduce the chances of creating hollows on the surface. Do not work any one area too much, and keep the sander moving.

PRACTICAL TIP

Where more material needs to be removed, this would normally be done with a belt sander.

STEP 5 Now sand each rail in turn until the whole rail has been sanded.

Figure 6.23 Sanding the rails, working along the grain

STEP 6 Repeat for the stiles.

STEP 7 Turn the door over and repeat Steps 3 to 5.

STEP 8 Fit a piece of 120 grit sandpaper into the sander.

STEP 9 Repeat Steps 5 and 6.

STEP 10 Clean down the sander after use and give it a visual check to ensure that it has not sustained any damage during use.

Store the sander and transformer in a purpose made box or the carrying case supplied with them, with the cable properly coiled. This in turn should be kept in a dry secure store.

PRACTICAL TASK

9. CUTTING OFF THE HORNS USING A PORTABLE POWERED CIRCULAR SAW

Perform your pre-use checks for the portable powered circular.

STEP 1 Measure the distance from the edge of the base to the edge of the saw blade (A). This will be different depending on which side of the blade you are working from. If you are cutting from right to left you will be using the wider side of the base, but if you are working from left to right you will be using the narrower side of the base.

Figure 6.24 Measuring the distance to the saw blade

STEP 2 Clamp a temporary guide across the door at the distance (A).

Figure 6.25 Clamping a temporary guide across the door

STEP 3 Set the depth of the cut by moving the base plate. Allow enough blade depth to allow the cut to be made without excessive blade protruding beneath the door.

Figure 6.26 Setting the depth of the cut (1)

Figure 6.27 The blade should not protrude too much beneath the door

STEP 4 Plug the circular saw into the mains supply via a 110V transformer.

STEP 5 Place the saw against the temporary fence and switch it on, making sure both hands are clear of the blade, and make the cut. Turn off the saw and allow the blade to stop before lifting from the work.

STEP 6 Repeat for the horns at the other end of the door.

STEP 7 Clean and store as for the other power tools.

USING CARPENTRY AND JOINERY PORTABLE POWER TOOLS TO DRILL AND INSERT FASTENINGS

Fastening and fixing to timber products can be carried out using a range of different types of screws, bolts, nails and adhesives. Each of the different types can be used for a variety of different surfaces, although some are more appropriate for different surfaces.

Plastic plugs

Plug fixings are used to attach timber to masonry. The plug is inserted into the hole that has been created using an electric drill. In the past carpenters used to make their own plugs out of timber, but nowadays it is far more common to use ready-made plastic plugs.

Figure 6.28 Plastic plugs are available in a variety of different sizes

Make sure that you choose a plastic plug that fits tightly into the hole you have drilled. The screw is then inserted through the timber and into the plastic plug. As the screw is tightened the segments that make up the plastic plug are pushed apart, giving it a very strong grip into the masonry.

Rawl bolts/expansion anchor bolts

Rawl bolts are useful if you are drilling into brickwork, stone or concrete. You should avoid drilling into mortar joints. The rawl bolt is basically an expanding, anchored fastener. You will need to drill a hole of the same diameter as the rawl bolt. The nut and washer is removed and the anchor is put into the hole. The timber is then fixed over the thread and then the washer and nut are fastened back onto the bolt.

Coach screws

A coach screw looks like a screw but has the head of a bolt. They are used for a wide variety of timber work in buildings. They are considered to be heavy duty. The shaft tapers to a point and as the screws are driven into the timber, using either a wrench, spanner or pliers, the head stops it from going all the way through. Coach screws are often used at major joints in timber. They are considered to be better than nails, mainly because they are stronger and can be removed easily if necessary.

Figure 6.29 A rawl bolt/expansion anchor bolt

Screws

Screws are graded according to their head type, their length and their gauge or diameter. Carpenters use screws with countersunk heads when the screw needs to be flush or below the surface of the material. Round-headed screws are used when sheet material is attached to timber. Raised head screws are normally used to attach metal fixings, such as door handles.

Screws also have different types of heads and there are screwdrivers or tools to match each type and size. The three common head types are Slotted, Phillips and Pozidriv. You may also come across Torx, Supredrive, Spax and Clutch.

Figure 6.30 A coach screw

Cavity fixings

Different sorts of fixings have been designed to cope with hollow walls.

There are a number of different options:

* Rubber sleeved fixing (Rawlnut) – a hole is drilled in the board and as the screw is tightened the rubber sleeve is compressed against the reverse side of the wall.

* Plastic collapsible fixing (Poly-Toggle) – this is a very similar method to the rubber sleeve, but instead uses a collapsing plastic sleeve. As the screw is tightened the plastic sleeve is pulled back on itself tightly against the inside of the board.

* Plastic spread fixing (plasterboard plug) – the plastic fixing is pushed into the drilled hole and an ordinary wood screw is tightened. The

Figure 6.31 Screws are available in a variety of sizes

fixing is pushed through the hole in the board. Small plastic legs of the fixing are then compressed against the back of the board to hold the screw in place.

* Spring metal toggle – these are stronger anchors and consist of a pair of spring-loaded metal arms with a thread tapped into a hinge pivot. The hole is drilled and the device is then pushed through the hole. Once the spring arms are free of the hole they will spread out onto the reverse of the board. The screw is then tightened, which compresses the arms against the back of the board.

* Gravity metal toggle – these work in a similar way to the spring metal toggle except that the toggle is only made of one piece of metal. It lies parallel to the screw as it is pushed through the drilled hole. Once it is through the hole, gravity takes over and the toggle drops to a vertical position. The screw is then tightened, which compresses the toggle against the back of the board.

Figure 6.32 Cavity wall fixings

Coach bolts

A coach bolt has a domed head with a square shank. A hole is drilled to accept the bolt and is tightened using a nut. They are often used for connecting timber trusses and are considered to be heavy duty and strong fixings.

Nails

Nails can be fixed into timber using either a hammer or a mechanical tool, such as a nail gun (you should never use a nail gun if you have not been trained to do so). Nails are either made from ferrous metal (this means they contain iron and could rust) or non-ferrous metal (which does not contain iron).

Figure 6.33 A coach bolt

There is a huge variety of different nails in terms of style, shape and size:

* Round wire – these are generally for first fixing work, or low quality jobs and tend not to be driven below the surface of the timber so they can easily be removed.

* Annular ring – these have a number of rings along their shank that gives them a stronger hold but at the same time makes them tougher to get out of the timber. They are covered in a thin layer of zinc to reduce the chance of them rusting.

* Lost head – these are nails that are used to fix floor boards. They can be driven in without the need to punch them below the surface due to the shape of the head.

* Oval wire – these are designed to go beneath the surface of the timber and because of their shape they allow the grain to close around the timber as they are punched below the surface.

* Panel pins – these are thin nails that are used to fix fine mouldings and beading. They are circular in cross-section and are designed to be punched below the surface of the timber.

* Cut nails and floor brads – cut nails are made from mild steel and are square in section. They give a good grip. They are used for fixing timber to blockwork. Floor brads are very similar to cut nails and are used for surface fixing of floorboards.

* Plasterboard nails – these are treated to prevent them from rusting and have a rough shank to give them a good grip. They are used to fix plasterboard or insulation board to joists and stud work.

* Wire clout – these are short nails with large heads and are used for fixing roof felt.

* Hardened steel – these are similar to lost head nails but are made of hardened zinc-plated steel. They are used mostly for fixing to brickwork and concrete.

* Duplex – these have double heads. The lower head is driven into the surface of the material and the upper head stays above the surface, making them easier to pull out. They are mainly used for temporary fixings.

* Machine-driven – these are nails and pins that are in coils or strips and used by pneumatic nailing machines. There is a huge range of different types and they can be used for structural work or final fixing of mouldings and trims.

Figure 6.34 Nails are available in many different sizes

Chemical fixing

There is a wide range of different adhesives that create a chemical bond, rather than relying on punching a hole through the material with a nail, screw or other type of fixing.

The adhesive always needs to penetrate the surface and key into the layers. This is a process known as mechanical adhesion.

The adhesive also has to have sufficient strength. This is why adhesives tend to be applied in a liquid state. As they harden, set or cure they become solid and stronger. This is achieved in a number of different ways:

* Solvent adhesive – when the adhesive is applied the solvent evaporates, or is absorbed into the timber.

* Cooling – a hot glue gun heats up the adhesive, transforming it from solid to liquid. It is applied in the liquid form but as it cools it hardens.

* Chemical – the adhesive needs a hardener or another chemical, or sometimes heat, to make it transform from a liquid to a solid. Synthetic resin adhesives are a good example. These are two-part powder or liquid adhesives.

* Combination – some adhesives use their loss of solvent, a chemical reaction and cooling at the same time to make them strong.

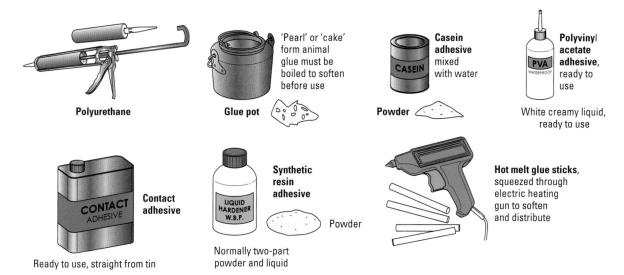

Figure 6.35 Types of adhesive

Safe use of chemical fixings

Always ensure that you read the instructions on the container of each chemical fixing you select for a task. This information will provide you with details of:

* its safe and effective use

* safety equipment required (PPE)

* correct storage and transportation

* what to do if you accidentally swallow or get the product on your skin or in your eyes.

Locating services

Whenever you are drilling or cutting there is a danger that just beneath the surface is either a cable or a pipe. It does not necessarily follow that water or gas pipes will only be found under floorboards. There also may be no logic behind the direction electric cables take inside walls. Not even cavity walls can be considered safe, as the cavity may have been used for electric cables and pipes. This is the same for partition walling inside a building.

There are devices that you can use to detect the flow of water, the presence of metal or an electric current.

PRACTICAL TIP

It is never a good idea to absolutely trust service detectors or locators. If you think you are going to have to drill or cut in an area where there might be services the wise precaution is always to turn those services off at the mains. This means that if any damage is done to pipes or cabling then there will not be an immediate danger and they can be repaired before the mains are switched back on.

TEST YOURSELF

1. What voltage of electrical tools should be used to minimise the risk of electric shock?

   a. 110V

   b. 150V

   c. 230V

   d. 240V

2. Which of the following tasks is made easier by a planer?

   a. Chamfering

   b. Rebating

   c. Edging

   d. All of these

3. What are the particles made of that make PCD drill bits especially strong?

   a. Carbon

   b. Tungsten

   c. Gold

   d. Diamond

4. What is a PAT test?

   a. A competency test to check you can use power tools safely

   b. An inspection made to an appliance that has broken

   c. A test for an appliance's safety before its first use

   d. A test to check if an appliance is still safe to use

5. When dealing with waste, elimination, reduction and re-use are all recommended. But which reduction and disposal method has been missed out?

   a. Landfill

   b. Recycle

   c. Repair

   d. Re-order

6. Which fixing device is useful if you are trying to fix to brickwork, stone or cement?

   a. Rawl bolt/expansion anchor bolt

   b. Coach screw

   c. Cavity fixing

   d. Nail

7. What is the term used to describe nails that have double heads?

   a. Annular ring

   b. Wire clout

   c. Duplex

   d. Machine driven

8. What feature do plasterboard nails have to give them a good grip?

   a. An enlarged tip

   b. A larger head

   c. A thicker diameter

   d. A rough shank

9. What metal is used to make cut nails and floor brads?

   a. Mild steel

   b. Iron

   c. Zinc

   d. Brass

10. Service locating machines can detect which of the following?

    a. Metal

    b. Water flow

    c. Electric current

    d. All of these

INDEX

ACKNOWLEDGEMENTS

The author and the publisher would also like to thank the following for permission to reproduce material:

Images and diagrams

Alamy: Arcaid Images: 3.12, blickwinkel: 3.13, Mike Booth: 4.71, Peter Davey: chapter 1 opener; **BSA**: 2.5; **Energy Saving Trust © 2013**: 3.28; **Fotolia**: 1.1, 1.2, 1.3, 1.5, 1.6, 1.7, 1.8, 1.14, 1.15, 1.16, 4.72; **Helfen**: 2.4; **instant art**: table 1.15; **iStockphoto**: 1.11, 2.26, 3.4, 3.16, 3.18, 3.19, 3.21, 3.22, 3.25, 3.26, 3.27, 4.33, 4.38, 4.39, 4.42, 4.43, 4.44, 4.47, 6.1; **KATRES**: 4.32; **Nelson Thornes**: 1.9, 1.10, 1.12, 1.13, chapter 4 opener, 4.1, 4.2, 4.3, 4.4, 4.5, 4.6, 4.8, 4.9, 4.10, 4.11, 4.12, 4.36, 4.37, 4.40, 4.41, 4.45, 4.46, 4.48, 4.49, 4.50, 4.51, 4.52, 4.53, 4.80, 4.83, 4.84, 4.86, 4.87, 4.88, 4.89, 4.90, 4.91, 4.92, 4.93, 4.94, 4.95, 4.96, 4.97, 4.98, 4.99, 4.100, 4.101, 4.102, 4.103, 4.104, 4.105, 4.106, 4.107, 4.108, 4.109, 4.110, 4.111, 4.112, 4.113, 4.114, 4.115, 4.116, 4.117, 4.118, 4.119, 4.120, 4.121, 4.122, 4.123, 4.124, 4.125, 4.126, 4.127, 4.128, 4.129, 4.130, 4.131, 4.132, 4.133, 4.134, 4.135, 4.136, 4.137, 4.138, 4.139, 4.140, 4.141, 4.142, 4.143, 4.144, 4.145, 4.146, 4.147, 4.148, 4.149, 4.150, 4.151, 4.152, 4.153, 4.154, 4.155, 4.156, 4.157, 4.158, 4.159, 4.160, 4.161, 4.162, 4.163, 4.164, 4.165, 4.166, 4.167, 4.168, 4.169, 4.170, 4.171, 4.172, 4.173, 4.174, 4.175, 4.176, 4.177, 4.178, 4.179, 4.180, 4.181, 4.182, 4.183, 4.184, 4.185, 4.186, 4.187, 4.188, 4.189, 4.190, 4.191, 4.192, chapter 5 opener, 5.18, 5.19, 5.20, 5.21, 5.22, 5.23, 5.24, 5.25, 5.26, 5.27, 5.28, 5.29, 5.30, 5.50, 5.51, 5.52, 5.53, 5.54, 5.55, 5.56, 5.57, 5.58, 5.59, 5.60, 5.61, 5.62, 5.63, 5.64, 5.65, 5.66, 5.67, 5.68, 5.69, 5.70, 5.71, 5.72, 5.73, 5.74, 5.75, 5.76, 5.77, 5.78, 5.79, 5.80, 5.81, 5.82, 5.83, 5.84, 8.85, 5.86, 5.87, 5.88, 5.89, 6.7, 6.8, 6.9, 6.10, 6.11, 6.12, 6.13, 6.14, 6.15, 6.16, 6.17, 6.18, 6.19, 6.20, 6.21, 6.22, 6.23, 6.24, 6.25, 6.26, 6.27, 6.29, 6.30, 6.31, 6.33, 6.34; **Peter Brett**: 2.1, 2.3, 2.6, 2.7; **Science Photo Library**: Peter Gardiner: 1.4; **Shutterstock**: chapter 2 and 3 opener, 3.11, 3.14, 3.20, 3.23, 3.24, 4.7, 4.34, 5.11, 5.12, chapter 6 opener, 6.2, 6.3, 6.28, 6.32; **Wikipedia**: 3.15.

Every effort has been made to trace the copyright holders but if any have been inadvertently overlooked the publisher will be pleased to make the necessary arrangements at the first opportunity.